Guide to Accidents at Work

Guide to Accidents at Work

Nigel Tomkins
Legal Consultant

Michael Humphreys
Barrister

Matthew Stockwell
Barrister

JORDANS

Published by
Jordan Publishing Limited
21 St Thomas Street
Bristol BS1 6JS

British Library Cataloguing-in-Publication Data

A catalogue record for this book is available from the British Library.

ISBN 978 1 84661 076 9

Typeset by Letterpart Ltd, Reigate, Surrey

Printed in Great Britain by Antony Rowe Limited, Chippenham, Wiltshire

FOREWORD

This excellent book should be on every practitioner's desk (not shelf) when they consider any accident at work case.

It is a book that practitioners of all levels of experience will find invaluable.

Now that the most useful law (from the claimant's perspective) comes to us by way of European Union Directives, brought into force in the United Kingdom by way of Regulations, there is a clear need for an authoritative yet readable book in this area. This is that book, so that, as the authors say, cases that may otherwise be lost in common law negligence will now succeed.

The guidance given to the law as it applies to injury at work is comprehensive yet clear with individual chapters on negligence; statutory duty; contributory negligence; investigation; evidence and procedure as well as chapters looking in depth at the key Regulations. It deals with all of the relevant authorities and I am delighted to warmly welcome this book and firmly recommend it to all personal injury practitioners.

Martin Bare
President, APIL

ASSOCIATION OF PERSONAL INJURY LAWYERS (APIL)

APIL is the UK's leading association of claimant personal injury lawyers, dedicated to protecting the rights of injured people.

Formed in 1990, APIL now represents over 5,000 solicitors, barristers, academics and students in the UK, Republic of Ireland and overseas.

APIL's objectives are:

- To promote full and just compensation for all types of personal injury;
- To promote and develop expertise in the practice of personal injury law;
- To promote wider redress for personal injury in the legal system;
- To campaign for improvements in personal injury law;
- To promote safety and alert the public to hazards;
- To provide a communication network for members.

APIL is a growing and influential forum pushing for law reform, and improvements, which will benefit victims of personal injury.

APIL has been running CPD training events, accredited by the Solicitors Regulation Authority and Bar Standards Board, for well over fifteen years and has a wealth of experience in developing the most practical up-to-date courses, delivered by eminent leading speakers, either publicly or in-house.

APIL training now runs almost 200 personal injury training events nationally each year, plus up to a further 100 meetings of our regional and special interest groups. Topics cover a wide range of subjects and are geared towards giving personal injury lawyers a thorough grounding in the core areas of personal injury law, whilst keeping lawyers thoroughly up-to-date in all subjects.

APIL is also an authoritative information source for personal injury lawyers, providing up-to-the minute PI bulletins, regular newsletters and publications, information databases and online services.

For further information contact:

APIL
11 Castle Quay
Nottingham
NG7 1FW

DX 716208 Nottingham 42

Tel: 0115 9580585
Email: mail@apil.org.uk
Website: www.apil.org.uk

PREFACE

Every employee has the basic right to return home safely at the end of his or her working day (or night as the case may be). Equally, every person who is injured in consequence of the negligence or breach of duty of another should have fair access to full compensation for his or her injuries and losses.

This book is written, as we have collectively spent many years practising, with both of these principles firmly at the centre of our consideration.

This publication is not the first, much less the last, word on the subject of employer's liability and we are of course are indebted to the many jurists, practitioners and other commentators who have contributed positively to the continuous development of this important area.

We hope to play our own small part in this process by assisting the busy practitioner with an accessible, reliable and practical guide to the effective preparation and presentation of such cases.

For a salesperson the secret of success is to know your product. Lawyers are sometimes considered to be salespeople. Like them lawyers assemble a package to present in court or on paper to obtain a result. Knowledge is the key to successful progress and resolution of a case. If this book helps in the acquisition of such knowledge, we shall have achieved our aim and it will be a success.

We believe that this is a practical practitioners' guide to the successful pursuit of personal injury litigation. The fact that it appears as an APIL publication indicates that all three authors have tended to act for claimants. In fact we have acted for defendants as well during our careers and this book is not blind to the fact that defendants should not settle unmeritorious claims.

Our aim is to provide clear, reasonable, practicable and sustainable answers to the central problems that confront the busy litigator in a manner that encourages two things. Firstly thought and consideration of what is needed to progress claims to a successful conclusion. Secondly thought and consideration of what is needed to recognise when a case will not succeed and act accordingly.

We would like to acknowledge the encouragement and friendship of our colleagues (past and present) and we offer particular thanks to our respective partners and families for their support and patience during the preparation of this work.

We also offer sincere thanks to Denise Kitchener, APIL Chief Executive, and Cenric Clement Evans of Hugh James without whose long and unstinting efforts to convince us that the book should be written it would never have happened. To Martin Bare of Morrish & Co, APIL President, for providing the Foreword to this publication and we gratefully acknowledge his long-standing contribution to the cause of helping injured people and their families.

The task of completing this book has been immeasurably eased by the continuous help, encouragement and close attention to detail of our publisher, Tony Hawitt, our editor, Julian Roskams, and Sally Harper for her quick and accurate typing of some parts of the manuscript. We are very much indebted to them all, but final responsibility for the contents, of course, rests with us.

The law is as stated at February 2008.

Nigel Tomkins, Michael Humphreys, Matthew Stockwell

28 February 2008

CONTENTS

Part C
Effective Case Preparation and Presentation

TABLE OF CASES

References are to paragraph numbers.

TABLE OF STATUTES

References are to paragraph numbers.

TABLE OF STATUTORY INSTRUMENTS

References are to paragraph numbers.

PART A

GENERAL PRINCIPLES

CHAPTER 1

INTRODUCTION

1.1 BASICS

Sometimes cases surface in articles or reports that leave you wondering how on earth the reported outcome was achieved. Closer examination frequently shows that the law was incorrectly applied and/or inadequately pleaded.

A good example is *Brazier v Dolphin Fairway Ltd*[1] in which the best efforts of the Court of Appeal could not help the claimant, John Brazier.

Brazier was lifting a wooden pallet from a stack of such pallets. The pallets were often piled on top of each other to over head height. Attempting to remove a pallet from the top of the pile was a difficult exercise because the six-foot square wooden pallets were heavy and bulky. The pallet in question had been placed on top of a pile by an unidentified fellow employee. As Brazier moved the pallet from the top of a pile to place it on the ground he stumbled with it, it twisted and the result was a hernia.

Other men did the same job. Each operator would lift down a new pallet about once every half hour so that, in the course of an hour, ten pallets would be lifted down by different workmen. In the course of a working day, a large number of pallets would be lifted down and brought in to use.

At first instance, the trial judge, said:[2]

> 'In order to prove negligence, the onus being on the claimant, he has to prove that this operation which caused his injury was foreseeably dangerous.'

The judge concluded that he failed to do that. He held that Brazier had failed to prove that the system of work in force had given rise to a foreseeable risk of injury. Brazier unsuccessfully appealed.

---

1 [2005] EWCA Civ 1469.

2 Quoted by Smith LJ at para 8.

The case was pleaded on the basis of negligence alone. Had the case been pleaded under the Manual Handling Operations Regulations 1992,[3] it seems inconceivable that the case could have failed since the manual handling operation of lifting down the pallets could well have been avoided. If not it would have been reasonably practicable to reduce the risk of injury by very simple steps, such as halving the maximum height of the stacks and having a system by which pallets were only lifted by two men not one .

Our intention is to try to help practitioners to avoid allowing the court to make decisions on the application of the law without first putting the correct (and binding) authorities before them. This should not be necessary but, unfortunately, there are many examples that show that it clearly is.

Our aim is to ensure that the John Braziers of this world win their cases.

1.2 SCOPE OF THE WORK

In the following chapters we shall look at all major aspects of the law on accidents at work. There are individual chapters on negligence; statutory duty; contributory negligence; investigation; evidence and procedure. This is in addition to a series of chapters looking in depth at some of the current key regulations on subjects such as the workplace, work equipment, manual handling and construction accidents.

1.3 INTERRELATIONSHIP BETWEEN DOMESTIC AND EUROPEAN LAW

The European basis of much of our current law goes all the way back to Art 118a of the Treaty of Rome 1957 which provides:

> 'Member States shall pay particular attention to encouraging improvements, especially in the working environment, as regards the health and safety of workers and shall set as their objective the harmonisation of conditions in this area, while maintaining the improvements made.'

Particular reference is made to the requirement to encourage improvement and also to maintain such improvement. Any law passed by Parliament must comply with the Treaty of Rome. Article 118a also provides that directives impose minimum requirements; this does not prevent member states improving on them. The fact that these are minimum requirements is also very important.

There is an impression that the UK 'gold plates' European Health and Safety Directives. The insurance industry claim to believe this is a fact, as

[3] SI 1992/2793.

do some employer's organisations. We believe that to a large extent this is a myth. The purpose of such European Directives is to improve the protection of the safety and health of workers at work.

The basics are simple. You start with the concept of and requirement to, carry out *risk assessments.* Regulation 3 of the Management of Health and Safety at Work Regulations 1999 requires the following:

> 'Every employer and self-employed person shall undertake a suitable and sufficient assessment of the risks to health and safety of employees and persons not in his employment.'

The regulations require employers to have arrangements in place to cover health and safety that should be integrated with management systems. If a risk is found it should, if possible, be eliminated. If it is impossible to eliminate a risk the Directives require the employer to *minimise* the risk.

The fact that those running companies owe a duty of care to those who may be affected by their activities is no different from the previous common law position. Regulation 3 is simply a modern application of Lord Atkin's famous ratioin *Donoghue v Stephenson* when he held that we all owed a duty of case to our neighbours defining them[4] as:

> 'Persons who are so closely and directly affected by my act that I ought reasonably to have them in contemplation as being so affected when I am directing my mind to the acts or omissions which are called into question.'

There has certainly been no increase in the duty as non-employees are still unable to sue for a breach of the management regulations.

So are these Directives important? In *Lewis v Avidan Ltd*[5] May LJ said:[6]

> 'These regulations were made to implement a European Framework Directive (Council Directive 89/391/EEC) of 12 June 1989 and a Workplace Directive (Council Directive 89/654/EEC) of 30 November 1989. There is, I think, nothing relevant to be derived from these directives except that they contain minimum requirements.'

From a simplistic viewpoint he is correct. They impose minimum standards. The standards they have replaced were often very old. But the old law sets the standard, not the Directives. We cannot have standards lower than the Directives, nor can we reduce existing standards to meet them.

---

4 [1932] All ER Rep 1 at 2.
5 [2005] EWCA Civ 670.
6 At para 2.

A good example is *Stark v The Post Office*[7] Postman Stark was riding his bike when part of the stirrup of the front brake broke in two, and one part lodged between the front wheel and it's forks. He was thrown off and seriously injured. The cause of the stirrup breaking was either metal fatigue or some manufacturing defect. The first instance judge found that the 'defect would not and could not have been discoverable on any routine inspection' – 'a perfectly rigorous examination would not have revealed this defect'. The relevant regulation in the case, implementing part of the Work Equipment Directive, used these words:

> 'Every employer shall ensure that work equipment is maintained in an efficient state, in efficient working order and in good repair.'

Waller LJ described the Directive thus:

> 'It seems to me that all that can be said of the Directive is that it sought to bring in minimum requirements. It positively encouraged more stringent requirements if they already existed, and there is nothing in the Directive to discourage a Member State in fulfilling its obligations under the Directives from imposing more stringent duties than the minimum required.'

In fact the case was not won on the basis of the Directive but on the strength of the old UK law. Two House of Lords decisions were key: *Galashiels Gas Co Ltd v Millar*[8] and *Hamilton v National Coal Board.*[9] From these and other cases the word 'maintained' imported strict liability. Once broken something is not maintained. That is UK law not EU.

It is important to remember the preamble to the Framework Directive on Health and Safety which states explicitly that:

> 'This Directive does not justify any reduction in levels of protection already achieved in individual Member States.'

All of this means that short of leaving the EU there is no circumstance that would allow health and safety law to be downgraded. The draftsman of the UK Regulations in issue in *Stark* had used language construed as imposing a strict obligation over many years in the context of the health and safety of employees. That language gave effect to the minimum obligations, but it also went further to comply with pre-existing standards.

However, few of our EU based regulations impose such strict liability. In many instances, words have been used such as 'reasonably practicable' which permit a defence including a defence based on cost. However where use of the words 'reasonably practicable' open the way to a defence based on cost, that defence is strictly limited by the requirement to demonstrate gross disproportion between cost and risk.

---

7 [2000] PIQR P105.
8 [1949] AC 275.
9 [1960] AC 633.

Once again this is not new and is not a result of European law. It is the same standard as that was set under the old common law. This tends to be the problem that employers are unable to get round.

1.4 ESTABLISHMENT OF MODERN CONCEPT OF EMPLOYER'S DUTY OF CARE

The modern duty of care owed by an employer to an employee was established in 1937 by the House of Lords in *Wilson and Clyde Coal Co Ltd v English*.[10]

On 27 March 1933, English was employed underground in a mine. At the end of the day shift, he was on one of the main haulage roads when the haulage plant was turned on. Before he could get out, he was crushed. It seems inconceivable today that these facts could lead to a contested action, let alone one that went all the way to the House of Lords.

Lord Macmillan held that the provision of a safe system of working was an obligation on the employer. He went on to say:[11]

> 'He cannot divest himself of this duty, though he may —and, if it involves technical management and he is not himself technically qualified, must perform it through the agency of an employee. It remains the [employer's] obligation and the agent whom the [employer] appoints to perform it, performs it on the [employer's] behalf. The [employer] remains vicariously responsible for the negligence of the person whom he has appointed to perform his obligation for him, and cannot escape liability by merely proving that he has appointed a competent agent. If the [employer's] duty has not been performed, no matter how competent the agent selected by the [employer] to perform it for him, the owner is responsible.'

As Lord Wright said[12] the employer's duty is:

> 'Personal to the employer, to take reasonable care for the safety of his workmen, whether the employer be an individual, a firm or a company and whether or not the employer takes any share in the conduct of the operations.'

In other words there is a personal non-delegable duty which is at the core of the law on employer's liability. This duty includes the provision of a safe place of work, a safe system of work, the provision of competent staff and proper equipment.

Work should be as safe as is reasonably practicable. That is the standard in UK law, regulations or common law. Although there are many health

---

10 [1937] 3 All ER 628.
11 At 639.
12 At 641.

and safety requirements in the regulations imposing absolute standards or standards governed by practicability alone, reasonable practicability is the conventional standard in negligence.

Reasonable practicability is also the threshold for the standard of care in the general duties of Health and Safety at Work etc Act 1974, ss 2-7 that give rise to criminal but not to civil sanctions despite the generality of their wording (s 47(1)). It is also the primary standard in the Regulations that are intended to implement the European health and safety Directives.

1.5 STATUTORY DUTY

Cases based on breach of statutory duty frequently impose strict liability. But they fall into two categories: true strict liability and strict liability with a defence of reasonable practicability. How it works is again not a result of European law. The key case is *Nimmo v Alexander Cowan & Sons Ltd.*[13]

During the course of his employment Nimmo was unloading bales of pulp. He stood on one of the bales to unload others. He fell due to the tipping of the bale, and was injured. Nimmo sued his employers for breach of statutory duty under s 29(1) of the Factories Act 1961. That stated that there 'shall, so far as is reasonably practicable, be provided and maintained safe means of access' to working places and 'every such place shall, so far as is reasonably practicable, be made and kept safe for any person working there'.

Nimmo did not allege that it was reasonably practicable for his employers to make the working place safe. The House of Lords held that, on the true construction of s 29 and in view of the fact that a criminal offence was created, the onus of proving that it was not reasonably practicable to make the working place safer than it was lay on the employers. That meant that it was not necessary for Nimmo to allege that it was reasonably practicable. The burden was reversed and Nimmo's employers had to show that they had taken all reasonably practicable steps to make the workplace where he worked safe.

There is plenty of evidence to suggest that those who claim that the United Kingdom 'gold plates' European health and safety Directives are simply wrong and do not understand UK or EU law.

It is important to make use of both old and new law to ensure the right outcome in cases. We hope this book will help readers to achieve that.

[13] 1967 SC (HL) 79.

CHAPTER 2

GENERAL PRINCIPLES OF NEGLIGENCE

2.1 BASIC DUTIES

As well as obligations under statute and regulations, employers continue to have a common law duty of care to their employees. Failure to fulfill that duty is negligence. For example, employers owe a specific duty to their employees to provide them with safe premises, independently of any duty they may owe to them under the health and safety legislation.

It was not until 1937, with the decision of the House of Lords in England in *Wilson and Clyde Coal Co Ltd v English*,[1] that the modern duty of care owed by an employer to an employee was established. It is worth noting that this is a Scottish case. Time and time again the law of the United Kingdom comes from Scotland. If it was not for the efforts of solicitors and advocates in Scotland workers, in particular, might well be much less well protected under the law.

On 27 March 1933 English was employed underground in a mine. At the end of the day shift, he was on one of the main haulage roads when the haulage plant was turned on and, before he could get out, he was crushed. It seems inconceivable today that these facts could lead to a contested action at all, let alone one that went all the way to the House of Lords.

Lord Macmillan held that the provision of a safe system of working was an obligation on the employer. He went on to say:

> 'He cannot divest himself of this duty, though he may —and, if it involves technical management and he is not himself technically qualified, must perform it through the agency of an employee. It remains the [employer's] obligation and the agent whom the [employer] appoints to perform it, performs it on the [employer's] behalf. The [employer] remains vicariously responsible for the negligence of the person whom he has appointed to perform his obligation for him, and cannot escape liability by merely proving that he has appointed a competent agent. If the [employer's] duty has not been performed, no matter how competent the agent selected by the [employer] to perform it for him, the owner is responsible.'

---

1 [1937] 3 All ER 628.

As Lord Wright said the employer's duty is:

> 'Personal to the employer, to take reasonable care for the safety of his workmen, whether the employer be an individual, a firm or a company and whether or not the employer takes any share in the conduct of the operations.'

In other words there is a personal non delegable duty which is at the core of the law on employer's liability. This duty included the provision of a safe place of work, a safe system of work, the provision of competent staff and proper equipment. Vicarious liability is at the heart of the duty and this aspect of the law is considered in some detail below.

In the High Court in the Republic of Ireland[2] the case of *Barclay v An Post & Murray*[3] McGuinness J considered the modern duty of care based on these principles. The case concerned an employee of the defendant company, who sustained back injury while delivering letters to low level letterboxes in June 1993.

Bradley reported the matter to his supervisor and subsequently attended the company doctor. The injury resulted in an absence of two months by Bradley from his work. His injury and consequent vulnerability were well known to his employers, yet in October 1993 he was dispatched on non-emergency overtime to deliver mail to a development where some 350 houses had low level letterboxes.

During the course of this delivery Bradley again suffered back injury. McGuinness J held that the defendant company's duty of care towards Bradley included a duty to ensure that, at least in the short term after his illness, he did not assume duties that would place undue and extraordinary strain on his back. Consequently, the defendant company was held liable as it did not properly discharge the employer's duty of care in the case of Bradley's second injury.

McGuinness J's approach to the employer's duty of care is consistent with the approach adopted in previous cases. These cases decided that an employer discharges his duty of care 'if he does what a reasonable or prudent employer would have done in the circumstances'. The employer's duty is therefore not an unlimited one and varies according to the employee's circumstances.

---

[2] The law in Ireland is based on the same approach as United Kingdom law and the application here should be identical as should the outcome of any case on a the same set of facts.

[3] (unreported) 7 July 1998, HC.

2.2 THE CORE STANDARD

Work should be as safe as is reasonably practicable. That is the standard in UK law. Although there are many health and safety requirements in the regulations imposing absolute standards or standards governed by practicability alone, reasonable practicability is the conventional standard in negligence. Reasonable practicability is also the threshold for the standard of care in the general duties of Health and Safety at Work etc Act 1974, ss 2-7 which give rise to criminal but not to civil sanctions despite the generality of their wording: s 47(1). It is also the primary standard in the Regulations that are intended to implement the European health and safety directives. Pre-existing standards cannot be diminished by the regulations flowing from the directives, as shown in *Stark v The Post Office*.[4] Pre-existing standards are accordingly still of some considerable importance.

The concept of reasonable practicability is a device primarily to establish (or not) liability for accidents that have occurred either in civil or criminal law. Reasonable practicability includes economic considerations. It is part of the mechanism for the allocation of costs of accidents and preventative measures in society.

Reasonable practicability today is a relatively successful device in resisting the crudest form of free market economics that would remove all 'burdens from business' and, in the field of health and safety, let workers carry the whole risk. It certainly appears nearer the European standard than the coarsest form of cost benefit analysis.

In *Morris v West Hartlepool Steam Navigation*,[5] Lord Reid set out the considerations involved when an employer carries out what is today described as a risk assessment.

A risk assessment now has to be undertaken in relation to all aspects of the employer's activities under the Management of Health and Safety at Work Regulations (1993 & 1999), reg 3 and the Framework Directive, Art 3(a) of. Lord Reid stated:

> 'It is the duty of the employer in considering whether some precaution should be taken against a foreseeable risk, to weigh, on the one hand, the magnitude of the risk, the likelihood of an accident occurring and the possible seriousness of the consequences if an accident does happen, and, on the other hand, the difficulty and expense and any other disadvantage of taking the precaution.'

---

4 [2000] ICR 1013, CA.

5 [1956] AC 522.

2.3 NEGLIGENCE

There are four factors to be considered when negligence is at issue: the likelihood of an injury, potential seriousness of the outcome, proportionality and finally the size of the employer.

2.3.1 Likelihood

The approximate frequency of the foreseeable event happening may not need to be very great to trigger the need to take precautions, especially if the consequences would be serious if it did occur.

In *Bolton v Stone*,[6] a single minor injury in the course of 30 years as the result of some six cricket balls being hit over a 17-foot fence onto a road was considered by the House of Lords. As Lord Reid said when considering the likelihood of injury: 'If cricket cannot be played on a ground without creating a substantial risk, then it should not be played there at all'. However there was no liability. Nevertheless the standard was set; as Lord Reid stated:

> 'If this appeal is allowed that does not in my judgment mean that in every case where cricket has been played on a ground for a number of years without accident or complaint those who organise matches there are safe to go on in reliance on past immunity. I would have reached a different conclusion if I had though that the risk here had been other than extremely small, because I do not think that a reasonable man considering the matter from the point of view of safety would or should disregard any risk unless it is extremely small.'

In *Hilder v Associated Portland Cement Manufacturers Ltd*,[7] a motor cyclist was killed as a result of being hit by a ball kicked by a boy playing in a field adjoining the highway. The evidence was that the football was frequently kicked by boys over a low wall into the road, and the owners were held to be liable. Frequency made the difference.

2.3.2 Seriousness of Outcome

The most serious risk is life itself. Where such a risk is present the duty to exercise reasonable care demands the expenditure of great expense and trouble, if there is no easier way of obviating that risk.

In *Marshall v Gotham Co Ltd*,[8] Lord Reid said:

---

6 [1951] 1 All ER 1078.
7 [1961] 1 WLR 1434.
8 [1954] AC 360.

> '... if a precaution is practicable it must be taken unless in the whole circumstances that would be unreasonable. And as men's lives may be at stake it should not lightly be held that to take a practicable precaution is unreasonable.'

In the *Marshall* case the foreseeable danger was 'a very rare one'. The precautions were very onerous and 'would not have afforded anything like complete protection against the danger, and their adoption would have had the disadvantage of giving a false sense of security'. So it was held not reasonable to have taken them. This is completely different from the situation when strict liability applies through statute.[9] However at common law it is about balance. In *Read v Lyons*,[10] Lord MacMillan held: 'The law in all cases exacts a degree of care commensurate with the risk created'.

Paris v Stepney Borough Council,[11] illustrates this very well. The employers were negligent not to supply goggles to a one-eyed workman that would not have been necessary to fully sighted worker. Why? Because the risk of blindness if there is an accident is much more serious than the risk of the loss of sight in one eye. *Paris* sets the standard for all employers ensuring that they have to take their employees as they find them and treat them as is appropriate when it comes to the standard of care they need to exercise. In *Paris* Lord Morton said:

> 'In considering generally the precautions which an employer ought to take for the protection of his workmen it must, in my view, be right to take into account both elements, the likelihood of an accident happening and the gravity of the consequences.
>
> I take as an example two occupations in which the risk of an accident taking place is exactly equal; if an accident does occur in the one occupation, the consequences to the workman will be comparatively trivial; if an accident occurs in the other occupation the consequences to the workman will be death or mutilation.
>
> Can it be said that the precautions which it is the duty of an employer to take for the safety of his workmen are exactly the same in each of these occupations? My Lords, that is not my view.
>
> I think that the more serious the damage which will happen if an accident occurs, the more thorough are the precautions which an employer must take.'

The balancing of the scales in health and safety law has an air of unreality for many reasons. First of all the scale of the injury may not reflect the potential consequences of the failure. An unfenced hole in the floor may pose equally the risk of a twisted ankle or a fall causing a fatal blow to the

---

9 See *Skinner v Scottish Ambulance Service* [2004] ScotCS 176 below.

10 [1947] AC156.

11 [1951] AC 367.

head. An escape of dangerous gas may be detected by a night-watchman who sustains a few days' nausea or it may kill and injure thousands, as happened at Bhopal.

The second point is that in most work situations the consequences of the failure to take precautions may be relatively easy to foresee, both in terms of scale and frequency. You might think that this means that the duty is easy to discharge. Think again. The law does not limit liability to precisely foreseeable injury.

In *Jolley v Sutton LBC*,[12] the claimant was injured when a small abandoned cabin cruiser, which had been left lying in the grounds of a block of flats owned by the council, fell on him as he lay underneath it while attempting to repair and paint it. The judge had held that the type of accident and injury that occurred was reasonably foreseeable. He had stated in general terms that the risk was that children, including those of the age of the claimant, would 'meddle with the boat at the risk of some physical injury'.

Before the Court of Appeal the council had conceded that they had been negligent in failing to remove the cruiser but that such negligence had only created a risk of children being injured by rotten planking giving way beneath them, and had not created a risk of an accident in the circumstances that in fact occurred. The Court of Appeal held that it was not reasonably foreseeable that an accident would occur as a result of the claimant deciding to work underneath the propped-up boat, nor could any reasonably similar accident have been foreseen. The short point on appeal to the House of Lords was whether the wider risk as found by the judge, which would include within its description the accident that actually happened, was reasonably foreseeable.

The Court of Appeal had never squarely addressed the question whether the judge's critical finding that the type of injury was reasonably foreseeable was open to him on the evidence. In the view of the House of Lords it was a view that was justified by the particular circumstances of the case and the Court of Appeal had not been entitled to disturb the judge's findings of fact. The risk of children receiving injury in some way other than by falling through the planks could not be described as different in kind from that which should have been foreseen, wholly unforeseeable or too remote.

The judge's broad description of the risk as being that children would 'meddle with the boat at the risk of some physical injury' was the correct one to adopt on the facts of this case. There had been a concession by the council that the Lords said showed that where there was a wider risk, and the council would have had to incur no additional expense to eliminate it.

12 [2000] 1 WLR 1082, HL.

That decision is fully in line with *Hughes v the Lord Advocate.*[13] A manhole in an Edinburgh street was opened under statutory powers for the purpose of maintaining underground telephone equipment. It was covered with a tent and, in the evening, left by the workmen unguarded but surrounded by warning paraffin lamps. An eight-year-old boy entered the tent and knocked or lowered one of the lamps into the hole. An explosion occurred causing him to fall into the hole and to be severely burned.

The Lords held that the workmen were in breach of a duty of care to safeguard the boy against this type of occurrence which, arising from a known source of danger – the lamp – was reasonably foreseeable, although that source of danger acted in an unpredictable way.

This is still sound law. It applies in employer's liability cases just as in other areas, as was confirmed again by the House of Lords in *Simmons v British Steel plc*[14] Christopher Simmons tripped, fell and struck his head on a metal stanchion at work. There was a severe impact, but fortunately he was wearing protective head gear so his head injury was not as serious as it might have been. Nevertheless he sustained a severe blow to the head. He was dazed and shaking, and developed a swelling on the right side of his head. This was accompanied by headaches, disturbance to his eyesight and suppuration from his right ear. The trial judge held that the accident was caused by the fault of the employers. He awarded Simmons damages of £3,000, with interest, for these physical injuries.

However, the consequences of the accident were not confined to the physical injuries for which the judge awarded damages. After the accident Simmons experienced an exacerbation of a pre-existing skin condition, and he developed a change in his personality that resulted in a severe depressive illness. The judge found that Simmons' pre-existing skin condition was exacerbated and that he was suffering from a depressive illness and a complete change in his personality. But he was not satisfied that Simmons had proved on a balance of probabilities that either of these consequences had been caused by the accident.

The Lords held that the fact that Simmons sustained physical injuries for which the employers had been found liable made it unnecessary to ask whether the psychiatric injury from which he had also been suffering was reasonably foreseeable; *Page v Smith*[15] applied.[16] The starting point was that Simmons was a primary victim and therefore the employer had to take him as they found him. This meant that the aggravation of his skin

---

[13] [1963] AC 837.

[14] [2004] UKHL 20.

[15] (1995) P & CR 329.

[16] In *Page v Smith* it was held to be sufficient that the defendant should have foreseen that his negligent driving might cause some physical injury. It did not matter than he could not have foreseen that the event which actually happened, namely a minor collision, would cause psychiatric injury.

condition and the anger that led to his depressive illness could both be assumed to fall within the scope of its liability so long as there was a causal connection between the symptoms and the accident.

On the judge's findings Simmons' anger was one of the things caused by the accident. Emotional reactions such as anger, distress or fear did not sound in damages, however, emotional reactions could lead to other conditions, both physical and psychiatric, for which damages could be awarded. The trial judge had failed to consider, as he should have done given his findings, whether Simmons' anger at the accident itself had materially contributed to the exacerbation of his skin condition.

There were several causes of his anger and it was enough that one of them arose from the fault of the employer. Simmons' anger at the happening of the accident could not be dismissed under the *de minimis* principle. On the evidence, it had made a material contribution to the development of the skin condition and the depressive illness that resulted from it. It followed that the causal connection was established and Simmons was entitled to payment of the full amount of damages: he recovered £498,221.77 plus interest.

2.3.3 Proportionality

The weighing exercise is not done with a view to seeing whether the scales just tip. In *Edwards v National Coal Board*[17] a colliery worker was killed as a result of a fall that occurred from the side of a travelling road. The accident was due to a latent defect, known as a glassy slant, in the side of the roadway. The evidence showed that the defendants' officials had never applied their minds to the question whether it was reasonably practicable to provide artificial support for this roadway or not, although more than half the existing roads were already artificially supported.

It was held that the defendants had failed to discharge the burden that lay on them under the Coal Mines Act 1911, ss 49 and 102(8), which imposed on the mine owner an absolute duty to make the roof and sides of every travelling road secure. He could only escape civil liability for breach of this duty if he could prove, the onus being on him, that it was not 'reasonably practicable' to avoid or prevent the breach. On this basis the deceased's widow's claim for damages under the Fatal Accidents Acts 1846 succeeded.

In his decision, what was said by Lord Asquith is just as relevant in negligence. He said this:

> 'Reasonably practicable' is a narrower term than "physically possible", and seems to me to imply that a computation must be made by the owner in which the quantum of risk is placed on one scale and the sacrifice involved

[17] [1949] 1 KB 704, CA.

> in the measures necessary for averting the risk (whether in money, time or trouble) is placed in the other, and that, if it be shown that there is a gross disproportion between them – the risk being insignificant in relation to the sacrifice – the defendants discharge the onus on them.'

They key here is gross disproportion. On a proper analysis, then, the test is that reasonable practicability requires a precaution to be taken unless the time, effort and expense of taking it in relation to the risk averted is grossly disproportionate.

2.3.4 Size of Employer

It has long been accepted that limited resources do not give rise to a lower standard of care except perhaps in very limited circumstances where the defendant is a public body. If the organisation does not have sufficient resources to undertake its activities safely it should not undertake them. For confirmation see both *PQ v Australian Red Cross Society*,[18] and *Voli v Inglewood Shire Council*.[19]

However, an even higher standard of care is required of organisations of substantial means. This was confirmed by Lord Reid in *British Railways Board v Herrington*.[20] Such employers might be expected to take more expensive precautions and may be obliged to exercise initiative in devising or securing safety precautions not available to the smaller firm.

In *Knox v Cammell Laird Shipbuilders Ltd*,[21] Simon Brown J held that shipbuilding employers: 'were no more careless of the health of their workforce than other such commercial employers up and down the country'. However that was no defence to them for failure to protect the workforce from lung disease caused by fumes. The defendant was:

> 'in the forefront of the industry, one of the two largest yards in the country and with an enormous workforce. It was for employers like them to take the initiative in these matters, to be ever alert to the problems of their men and resourceful as to their solution.'

In *Allen v British Rail Engineering Ltd*,[22] Janet Smith J held that BREL:

> 'were a large employer and might have been expected to bring considerable pressure to bear on manufacturers who valued their custom.'

---

[18] [1992] 1 VR 19.
[19] (1963) 110 CLR 74.
[20] [1972] AC 877.
[21] (unreported) 30 July 1990, HC.
[22] (unreported) 7 October 1998, HC.

This meant that they should have 'encouraged' them to manufacture tools that gave off less injurious vibration than those that had caused the plaintiffs to develop vibration white finger. In a similar case at lower level this was followed.

It goes even further. In *Armstrong v British Coal Corpn*[23] the Court of Appeal held, affirming the judgment of HHJ Stephenson of 15 January 1996, that a large employer with vast resources was to be expected to research into conditions such as VWF from which its employees might be suffering.

The employer must not only incur the costs of the precautions indicated by its actual state of knowledge, it must incur the trouble and expense of keeping reasonably up to date with the knowledge of those precautions. For confirmation see both *Wright v Dunlop Rubber Co*[24] and *Wallhead v Rustons and Hornsby*.[25]

The key case is *Stokes v Guest Keen and Nettlefold Bolts & Nuts Ltd*.[26] The claimant's husband died from scrotal cancer that, it was established on the balance of probabilities, was induced by his exposure to contact with mineral oil during the 15 years he was employed as a tool-setter by GKN. Medical evidence was given to the effect that it took at least five years of regular exposure to produce an oil wart that might or might not become malignant and that some workers never developed the condition even after a lifetime of exposure to the risk.

From 1941 GKN employed a full-time medical officer who was responsible for the general safety organisation of the factory and had special knowledge of occupational medicine and industrial hygiene. From 1941 onwards medical scientists made recommendations that there should be periodical medical inspections of workers exposed to the risk of cancer and specific warnings given. In 1960 the factory inspectorate issued a leaflet describing warts on the scrotum as being potentially cancerous and recommending workers to have periodical medical checks.

GKN's medical officer considered that periodic examinations were out of proportion to the risk having regard to the low incidence of the disease and that warnings of the risk of cancer might frighten workers, with the result that warnings were not given and the leaflet not then circulated. After another tool-setter had died from scrotal cancer in 1963 the medical officer gave a talk on cancer to the factory works council, but it was unlikely that any warnings of symptoms ever reached the claimant's husband.

---

23 [1998] JPIL 320.
24 (1973) 13 KIR 255.
25 (1973) 14 KIR 285.
26 [1968] 1 WLR 1776.

In *Stokes* Swanwick J laid out, in a passage that has been repeated over and over again, what the standard of care demanded of an employer is:

> 'The overall test is still the conduct of the reasonable and prudent employer, taking positive thought for the safety of his workers in the light of what he knows or ought to know; where there is a recognised and general practice which has been followed for a substantial period in similar circumstances without mishap, he is entitled to follow it, unless in the light of common-sense or newer knowledge it is clearly bad; but, where there is developing knowledge, he must keep reasonably abreast of it and not be too slow to apply it; and where he has in fact greater than average knowledge of the risks, he may be thereby obliged to take more than the average or standard precautions.
>
> He must weigh up the risk in terms of the likelihood of injury occurring and the potential consequences if it does; and he must balance against this the probable effectiveness of the precautions that can be taken to meet it and the expense and inconvenience they involve. If he is found to have fallen below the standard to be properly expected of a reasonable and prudent employer in these respects, he is negligent.'

The employers were liable. That is still the standard to be applied today.

2.4 VICARIOUS LIABILITY

The most common use of negligence as a means of winning an employer's liability case is vicarious liability. It should be pleaded at every opportunity in addition to other allegations of negligence and breach of statutory duty.

The scope of vicariously liability is wide. For example it is now clearly established that intentional torts, including deliberate sexual abuse, are not inconsistent with vicarious liability by the decision of the House of Lords in *Lister v Hesley Hall Ltd.*[27]

In 1979 Axeholme House, a boarding annex of Wilsic Hall School, Wadsworth, Doncaster, was opened. Between 1979 and 1982 the claimants were resident at Axeholme House. At that time the claimants were aged between 12 and 15 years. The school and boarding annex were owned and managed by Hesley Hall Ltd as a commercial enterprise. In the main, children with emotional and behavioural difficulties were sent to the school by local authorities. Axeholme House was about two miles from the school.

The aim was that Axeholme House would provide care to enable the boys to adjust to normal living. It usually accommodated about 18 boys. The company employed Mr and Mrs Grain as warden and housekeeper to

---

[27] [2002] 1 AC 215.

take care of the boys. The employers accepted that at the material time they were aware of the opportunities for sexual abuse that might present themselves in a boarding school environment.

The warden was responsible for the day-to-day running of Axeholme House and for maintaining discipline. He lived there with his wife, who was disabled. On most days he and his wife were the only members of staff on the premises. He supervised the boys when they were not at school. His duties included making sure the boys went to bed at night, got up in the morning and got to and from school. He issued pocket money, organised weekend leave and evening activities, and supervised other staff. Axeholme House was intended to be a home for the boys, and not an extension of the school environment.

The employers accepted that, unbeknown to them, the warden systematically sexually abused the claimants in Axeholme House. The sexual abuse took the form of mutual masturbation, oral sex and sometimes buggery. The sexual abuse was preceded by 'grooming', being conduct on the part of the warden to establish control over the claimants. It involved unwarranted gifts, trips alone with the boys, undeserved leniency, allowing the watching of violent and x-rated videos, and so on. What might initially have been regarded as signs of a relaxed approach to discipline gradually developed into blatant sexual abuse. Neither of the claimants made any complaint at the time.

In 1982 the warden and his wife left the employ of the defendants. In the early 1990s a police investigation led to criminal charges in the Crown Court. Grain was sentenced to seven years' imprisonment for multiple offences involving sexual abuse. In 1997 the claimants brought claims for personal injury against the employers.

The question for the House of Lords was whether the warden's torts were so closely connected with his employment that it would be fair and just to hold the employers vicariously liable. The Lords applied the test laid down in *Salmond and Heutson on Torts*.[28]

The passage was originally drafted before the decision in *Lloyd v Grace, Smith & Co*[29] that affirmed that vicarious liability could still arise where the fraud of an agent was committed solely for the benefit of the agent. But as Lord Clyde said it has remained as a classic statement of the concept:

> 'A master is not responsible for a wrongful act done by his servant unless it is done in the course of his employment. It is deemed to be so done if it is either (1) a wrongful act authorised by the master, or (2) a wrongful and unauthorised mode of doing some act authorised by the master.'

---

28 (Sweet & Maxwell, 21st edn, 1996) p 443.

29 [1912] AC 716.

On the second of these two cases the text continues:

> 'But a master, as opposed to the employer of an independent contractor, is liable even for acts which he has not authorised, provided they are so connected with acts which he has authorised that they may rightly be regarded as modes – although improper modes – of doing them.'

The Lords said that in applying the *Salmond* test it was crucial to focus on the right act of the employee and its connection with the tortious act. The court must not simply consider whether the acts of sexual abuse were modes of doing an authorised act but must also consider whether there existed a close connection between the tort and the employee's duties.

In this case they held that the employer had undertaken to care for the resident children and had entrusted that obligation to the warden. Was there vicarious liability? On the facts of the case the answer to the question was yes. The court held that the sexual abuse was inextricably interwoven with the carrying out by the warden of his duties in Axeholme House. It clearly fell on the side of vicarious liability. They concluded that the warden's torts were so closely connected with his employment that it was fair and just to hold the employer vicariously liable.

Other cases back this approach up. The essential test for an employer's vicarious liability for the intentional torts of an employee is that of close connection, as formulated by Lord Nicholls in *Dubai Aluminium Co Ltd v Salaam*.[30]

The decision of the Court of Appeal in *Fennelly v Connex South Eastern Ltd*[31] is a good example. A ticket inspector checked Mr Fennelly's ticket that was shown with some reluctance. There was an altercation during which the ticket inspector assaulted Fennelly.

The Court of Appeal held that it followed immediately from the inspector carrying out the duties of his employment. It was artificial to say that the assault was divorced from what he was employed to do, which was to challenge whether Fennelly had a ticket, on behalf of his employer and the series of events comprised a single incident. They were vicariously liable as there was nothing to suggest that the inspector carried out the assault for his own purposes.

There was a similar outcome in *Mattis v Pollock (T/A Flamingo's Nightclub)*.[32] In 1998 Mattis was stabbed by a doorman employed at a nightclub owned and operated by Pollock. As a result Mattis was left paraplegic. The doorman was convicted of causing grievous bodily harm

---

30 [2002] UKHL 48.

31 [2001] IRLR 390.

32 [2003] EWCA Civ 887.

with intent and ordered to serve eight years in prison. He was not licensed as a doorman by the local authority.

On the night in question the doorman started a fight in the club in which Mattis and others were involved. The doorman then left the club, returned to his own flat nearby and armed himself with a knife. He returned to the vicinity of the club and stabbed Mattis who stood his ground outside the club.

The Court of Appeal held that Pollock was vicariously liable for the doorman's attack as it was so closely connected with what Pollock authorised or expected the doorman to do in the performance of his employment as a doorman that it would be fair and just to impose liability.

In a similar case, *Hawley v Luminar Leisure plc*,[33] the claimant was assaulted by a door supervisor not employed by the owners of the premises but employed by a firm contracted to provide to the owners security services in the form of door stewards at premises in Southend.

Who was vicariously liable? The principles were expressed perfectly in the case of *Denham v Midland Employers Mutual Assurance Ltd*[34] by Denning LJ (as he then was) when he said:

> 'The real basis of the liability is, however, simply this: if a temporary employer has the right to control the manner in which a labourer does his work, so as to be able to tell him the right way or the wrong way to do it, then he should be responsible when he does it in the wrong way as well as the right way. The right of control carries with it the burden of responsibility.'

The court recognised that Luminar sought to have and did exercise detailed control, not only over what the door stewards were to do in supplying services, but how they were to do it. That control was such as to make them temporary deemed employees of Luminar for the purposes of vicarious liability and they were vicariously liable for the conduct of the door supervisor.

Larking about can attract vicarious liability too. In *Harrison v Michelin Tyre Co Ltd*[35] Harrison was injured at work while standing on a duckboard of his machine talking to a fellow employee. Another employee while pushing a truck along a passage in front of Harrison decided as a joke to suddenly turn the truck two inches outside the chalked lines of the passageway and push the edge of it under Harrison's duckboard. The duckboard tipped up and Harrison fell off suffering injury.

---

33 [2006] EWCA Civ 18.
34 [1955] 2 QB 437.
35 [1985] 1 All ER 918.

It was defended on the basis that the employee had embarked on a frolic of his own. Comyn J held that the test for determining vicarious liability was whether a reasonable man would say either that the employee's act was part and parcel of his employment (in the sense of being incidental to it) even though it was unauthorised or prohibited by the employer, in which case the employer was liable, or that it was so divergent from his employment as to be plainly alien to it and wholly distinguish from it, in which case the employer was not liable. The employer was vicariously liable for their employee's negligence.

However it is not always that simple. In *McCready v Securicor*[36] the plaintiff and a colleague, both employees of Securicor, were playing on trolleys used at their work. While doing so, the plaintiff's hand was crushed when the other employee closed a vault door on it. The appeal court held that an employer was not vicariously liable for his employee's negligence where the employee's unauthorised and wrongful act was independent of any authorised act.

They held that 'in the course of his employment' and 'within the scope of his employment' have the same meaning; and the other employee was not acting for his employer when he closed the door on the plaintiff's hand, but was engaged in a prank.

To understand the scope of the employer's liability it is worth remember what Lord Millett said at paragraph 79 in *Lister v Hesley Hall Ltd*:

> 'So it is no answer to say that the employee was guilty of intentional wrong doing, or that his act was not merely tortious but criminal or that he was acting exclusively for own his own benefit, or that he was acting contrary to express instructions, or that his conduct was the very negation of his employer's duty.'

Overall however it is clear that it is very difficult for the employer to escape vicarious liability.

2.5 CAUSATION AND REMOTENESS

It is trite law that the burden of proving the link between the breach of duty and the injury is on the claimant.

'The judge made findings of fact and we cannot interfere . . .' is the straight bat often applied in appeal cases to uphold the judgment of the trial judge. Occasionally the appellate court will express surprise or the view that they would or might have decided the matter differently, but the findings must remain and the appeal is dismissed.

---

[36] [1991] NI 229.

Causation is a question of fact and, except in cases of perversity, first instance decisions will not be overturned. Therefore it is essential to understand that causation is at the heart of every case and it is therefore necessary to understand the legal meaning of causation. The principles governing causation and remoteness in respect of common law negligence apply equally to cases involving breach of statutory duty (for which see chapter 3 below).[37]

2.5.1 The Basic Concepts – 'but for' and 'material contribution'

In *Stapley v Gypsum Mines*[38] Lord Reid put it this way:

> 'To determine who caused an accident from the point of view of legal liability is a most difficult task. If there is any valid logical or scientific theory of causation it is quite irrelevant in this context... The question must be determined by applying common sense to the facts of each particular case, one may find as a matter of history, that several people have been at fault and that if any one of them had acted properly the accident would not have happened, but that does not mean that the accident must be regarded as having been caused by the faults of all of them. One must discriminate between those faults which must be discarded as too remote and those which must not.
>
> Sometimes it is proper to discard all but one parties fault and regard that one as the sole cause, but in others it is proper to regard two or more as having jointly caused the accident.'

The basic test of causation is usually expressed in the shorthand form as the 'but for' test. At first glance it might indicate that the injury was sustained because of a single event. However, matters are often less straightforward. In *Clough v First Choice Holidays and Flights Ltd*[39] Phillips LJ said that the term 'but for',

> '[...] encapsulate a principle understood by lawyers but applied literally or as if the words embody the entire principle the words can mislead. The Claimant is required to establish a causal link between the negligence of the Defendant and his injuries, or in short, that his injuries were indeed consequent on the negligence. Although on its own it is not enough for him to show that the Defendant created an increased risk of injury the necessary casual link would be established if, as a matter of inference from the evidence, the Defendant's negligence made a *material contribution* to the Claimant's injuries.'

---

37 *Fairchild v Glenhaven Funeral Services Ltd Fox v Spousal (Midlands) Ltd Matthews v Associated Portland Cement Manufacturers (1978) Ltd* [2002] UKHL 22, [2002] 3 All ER 305, [2003] 1 AC 32, HL.

38 [1953] AC 663.

39 [2006] EWCA Civ 15.

Material contribution was defined by Lord Reid in Bonnington Castings Ltd v Wardlaw[40] as follows:

> 'What is a material contribution must be a question of degree. A contribution which comes within the exception de minimis non curat is not material but I think that any contribution that does not fall within that exception is material.'

In relation to accidents at work, common sense must be applied to determine the material events that led to the injury. In common with many legal principles the devil is in the detail.

2.5.2 Exception – extended concept of causation

The concept of causation is a device used by the courts to define the ambit of the consequences of a breach of duty. The traditional approach, set out above, has been found wanting in industrial disease cases. In *Fairchild*[41] the House of Lords allowed, as an exceptional circumstance, a lower level of proof whereby the threshold is for the claimant to prove that the exposure has materially contributed to the risk of injury – rather than proving a material contribution to the injury.

In *Clough*, the Court of Appeal rejected an argument that the exceptional criteria in *Fairchild* could be used in accident cases. It is important to keep the basic and extended causation principles separate because they are distinct and the difference is important. If the test in *Fairchild* was applied in accident cases then liability would be established in every case where there was a breach of duty.

2.5.3 Practical Application

When investigating a claim it is important to sift out those matters which are relevant and those which are not. A simple example of a slipping/tripping case illustrates the problem.

On the day of the accident the claimant arrived at work and six hours later his injury occurred. Intuitively, it will be understood that the claimant's arrival at work was not a legal cause of his accident. However, during the next six hours events will occur which lead to the incident that causes injury. It is necessary therefore, in most cases, to look at more immediate circumstances that lead to the accident.

In a slipping/tripping case the first part of the investigation will be concerned with whether or not there was something about the floor, or on the floor, that could have caused the injury. In the most straightforward of cases the claimant may just fall down and there would be soiling on the

40 [1956] AC 613.

41 [2002] UKHL 22.

clothing or the shoes. Alternatively, there may be a skid mark through the slippery substance. In such a case it will not be difficult to infer that the claimant slipped as a result of what was on the floor. In a tripping case the worker may feel the impact of the foot on the defect in the floor – that is evidence that the defect caused the fall; however, he may feel nothing but here the presence of the defect should be enough to prove that the fall was caused by the breach of duty.

However, the matter may become more complicated. The degree of soiling on the floor may have been very slight and there may be no physical evidence to show what caused the claimant to fall onto the floor. Here, evidence from other workers, of the presence of a substance on the floor, would be important. If necessary, in such a case, it may be necessary to collect evidence about previous accidents or previous spillages. While such evidence does not go directly to showing there was something on the floor at the time it may be used to infer that there was.

Alternatively, the case may be complicated because the client is rendered unconscious as a result of the fall, or is simply in too much pain to deal with collecting evidence (for you) before being taking for treatment. Here, documentary evidence may reveal what happened.

2.5.4 Employees' Conduct

It is worrying that the Court of Appeal[42] had to reaffirm, relatively recently, the basic premise that unless the employees conduct completely overrides the breach by the employer that the claimant must receive some compensation.[43] The employees conduct is usually a matter for contributory negligence, discussed in detail in chapter 5.

2.5.5 Remoteness

Causation and remoteness are often coupled together in arguments over the ambit of the consequences of a breach of duty. However, breach of statutory duty is defined by the wording of the act or regulation and therefore remoteness is not an issue that arises in respect of the statutory obligations considered in this book.

2.5.6 Control

A number of the regulatory regimes impose liability on the basis of who has control.[44] In some cases there may be a number of participants and the role of each will need to be examined to determine who did what and

---

42 *Anderson v Newham College* [2002] EWCA Civ 505.

43 See *Boyle v Kodak* discussed at para **10.5.2**.

44 See for example the Construction (Design and Management) Regulations 2007, SI 2007/320.

who had the legal duty to comply with the regulations. In such cases it is necessary to look beyond who caused the injury and look at where the legal responsibility rests even if that person has had no active physical role in the events leading to the injury.

CHAPTER 3

STATUTORY DUTY

3.1 EUROPE

The European Union is the source of all modern UK health and safety law. It produces directives that in the United Kingdom are given effect by health and safety regulations. The regulations are enforced by the Health and Safety Executive and the courts where they are interpreted by judges. Interpretation must be purposive. The judge must look behind the regulations to the original directive to see what its purpose was and he must then apply the law on that basis.

3.2 PURPOSIVE INTERPRETATION

The interpretation of the regulations has to be approached with that in mind, asking what was the purpose. General guidance was given by the European Court of Justice (in such cases as *Von Colson v Land Nordheim-Westfalen*[1]) and by national courts, both in the United Kingdom (eg by the House of Lords in *Litster v Forth Dry Dock & Engineering Co Ltd*[2]) and in other jurisdictions within the European

---

1 [1984] ECR 1891.

2 *Litster v Forth Dry Dock and Engineering Co Ltd* [1989] 2 WLR 634, HL Forth Dry Dock, went into receivership in September 1983. On 6 February 1984 the company and its receivers entered into an agreement with another company, Forth Estuary, for the transfer to Forth Estuary of certain business interests of Forth Dry Dock with effect from Forth Dry Dock's close of business that day at 4.30 pm. Later the same day, at about 3.30 pm, the receivers dismissed the workforce of the company. Twelve of the dismissed employees made a complaint of unfair dismissal to the industrial tribunal against Forth Dry Dock, and Forth Estuary were subsequently sited as a party to the proceedings. The industrial tribunal held that the employees had been unfairly dismissed by Forth Dry Dock and that Forth Estuary were liable to pay them compensation. Forth Estuary appealed to the employment appeal tribunal, who dismissed the appeal except in relation to the amount of compensation payable. Forth Estuary then appealed to the Court of Session. The Court of Session allowed the appeal ([1988] IRLR 289) on the basis that because the workforce had been dismissed one hour before the transfer of the business, the employees had not been employed 'immediately before' the transfer within the meaning of reg 5(3) of the Transfer of Undertakings (Protection of Employment) Regulations 1981. The 12 employees appealed to the House of Lords. The Lords held that the 1981 Regulations had been enacted for the purpose of complying with Council Directive 77/187 which provided for the safeguarding of employees' rights on the transfer of a business, that the courts of the United Kingdom were under a duty to

Union, and also of course in the light of any specific guidance available from any of these sources as to the interpretation of these particular directives.

This approach was well illustrated last year in *Dugmore v Swansea NHS Trust.*[3] One of the most striking elements in the judgment is the court's recognition that the COSHH regulations implement European directives, in particular Council Directive 80/1107/EEC and 88/364/EEC and that neither of these directives has anything to say about the civil liability of employers towards their employees, nor do they impose obligations directly comparable to the regulation they were considering. Lady Justice Hale said:

> 'Their purpose is expressly preventive. According to the 1980 preamble, the measures taken by Member States to protect workers from the risks related to exposure to chemical, physical and biological agents at work were to be approximated and improved; that protection 'should so far as possible be ensured by measures to prevent exposure or keep it at as low a level as is reasonably practicable'; the sorts of measures involved are limiting or even banning the use of certain agents, suitable working procedures and methods, hygiene, information and warnings for workers, surveillance of their health, keeping updated records of exposure and medical records.
>
> This all reinforces the view taken by Lord Nimmo Smith [two decisions of Lord Nimmo[4] Smith in the Outer House of the Court of Session were considered and are referred to below] that the purpose of the regulations is protective and preventive: they do not rely simply on criminal sanctions or civil liability after the event to induce good practice.
>
> They involve positive obligations to seek out the risks and take precautions against them. It is by no means incompatible with their purpose that an

---

follow the practice of the European Court of Justice by giving a purposive construction to directives and regulations issued for the purpose of complying with directives and that reg 5(3) had to be construed on the footing that it applied to a person employed immediately before the transfer or who would have been so employed if he had not been unfairly dismissed before the transfer for reason connected with the transfer; and the appeal was allowed.

3 [2002] EWCA Civ 1689.

4 *Nimmo v Alexander Cowan & Sons Ltd* [1968] AC 107, HL. During the course of his employment Nimmo was unloading bales of pulp. He stood on one of the bales for the purpose of unloading others, fell due to the tipping of the bale, and was injured. He sued his employers, Alexander Cowan & Sons Ltd, for breach of statutory duty under the Factories Act 1961, s 29(1) which provides that there 'shall, so far as is reasonably practicable, be provided and maintained safe means of access' to working places and 'every such place shall, so far as is reasonably practicable, be made and kept safe for any person working there.' Nimmo did not allege that it was reasonably practicable for the respondents to make the working place safe. Their lordships held that, on the true construction of s. 29 and in view of the fact that a criminal offence was created, the onus of proving that it was not reasonably practicable to make the working place safer than it was lay on the employer; and accordingly that it was not necessary for the claimant to allege that it was reasonably practicable.

> employer who fails to discover a risk or rates it so low that he takes no precautions against it should nevertheless be liable to the employee who suffers as a result.'

Another example of purposive interpretation can be seen by what Smith LJ said at para 35 in the Court of Appeal's judgment in *Allison v London Underground Ltd*:[5]

> 'The preamble to the Framework Directive states that the Treaty (of European Union) provides that 'the Council shall adopt, by means of Directives, minimum requirements for encouraging improvements, especially in the working environment, to guarantee a better level of protection of the safety and health of workers.
>
> The preamble expressly recognises that standards of worker protection vary between Member States. It makes plain that the Directive does not justify any reduction in the levels of protection already achieved in individual Member States. However, the preamble also says that the improvement of workers' health and safety should not be subordinated to 'purely economic considerations.
>
> It states that employers shall be obliged to keep themselves informed of the 'latest advances in technology and scientific findings concerning workplace design' ... so as to be able 'to guarantee a better level of protection of workers' health and safety'. The general tenor of the preamble is that the Directive seeks to achieve improvement, in line with recent advances in technology but not to achieve perfection.'

Although Smith LJ confirms her view that the requirement to encourage improvements does not in any way suggest that member states will be expected to impose no-fault liability this passage provides a reminder of how important purposive interpretation is.

Finally, it is worth bearing in mind that where the defendant is such as a local authority, and hence an emanation of the state that directives can be relied upon directly against the state in the event that they are capable of direct effect.[6]

3.3 NEGLIGENCE VERSUS STATUTORY DUTY?

It is well worth comparing the old common law standards with modern statutory duty. In *Skinner v Scottish Ambulance Service*[7] the appellate court was considering a 'needlestick' injury to the pursuer's left thumb.

His case was that the injury would have been avoided, or that it would have greatly reduced the risk of injury of the type which happened to him

---

5 [2008] EWCA Civ 71.

6 See *Fratelli Costanzo SpA v Commune di Milano* [1989] ECR 1839.

7 [2004] ScotCS 176.

if a safer and more expensive type of canula had been used. The defender's answer was that any alternative would still present a risk to employees and would lead to 'substantially higher costs'.

The statutory case was based on reg 4 of the Provision and Use of Work Equipment Regulations 1998 [8]which provides that:

> '(1) Every employer shall ensure that work equipment is so constructed or adapted as to be suitable for the purpose for which it is used or provided.
>
> (2) In selecting work equipment, every employer shall have regard to the working conditions and to the risks to the health and safety of persons which exist in the premises or undertaking in which that work equipment is to be used and any additional risk posed by the use of that work equipment.
>
> (3) Every employer shall ensure that work equipment is used only for operations for which, and under conditions for which, it is suitable.
>
> (4) In this regulation 'suitable' … means suitable in any respect which it is reasonably foreseeable will affect the health or safety or any person.'

Skinner's argument was that on a proper construction of reg 4 and, in particular, reg 4(4)(a), it imposed on employers what was in effect an absolute requirement that work equipment should eliminate any reasonably foreseeable risk to the health or safety of an employee. A piece of work equipment either came up to that standard or it did not, but the question of cost was, in any event, irrelevant.

The employer's argument was that there was nothing in reg 4 or, indeed, elsewhere in the regulations or Council directives lying behind them that justified the construction contended for by Skinner. They said that the word 'suitable' should simply be given its ordinary broad meaning such as would allow in cost as being a possibly relevant factor.

In reality the question of relative costs was in a sense secondary to a more fundamental divide between the parties, namely whether, on a proper construction, reg 4 does or does not impose on an employer the sort of unqualified obligation (albeit limited by reasonable foresight) contended for on behalf Mr Skinner. As the court recognised if it did, then the question of relative costs became irrelevant because the defenders, in effect, conceded a breach of the regulation on that interpretation.

The legal test of reasonable practicability was looked at and compared with the test of practicability. The court confirmed that the test of reasonable practicability permits consideration of gross disproportion between cost and risk. They recognised that the legal significance of these tests has been established by judicial precedent going back many decades.

---

[8] SI 1998/2306.

As they noted the test of reasonable practicability permits consideration of gross disproportion between cost and risk.[9] The test of practicability does not. *A fortiori*, where the employer's obligation is couched in unqualified terms, no consideration of cost is admissible.

In *Summers v Frost*[10] Viscount Simonds observed:

> 'First, it appears to be an illegitimate method of interpretation of a statute, whose dominant purpose is to protect the workman, to introduce by implication words of which the effect must be to reduce that protection.
>
> Second, where it has been thought desirable to introduce such qualifying words, the legislature has found no difficulty in doing so …
>
> Third, it was decided as long ago as 1919 … that, if the result of a machine being securely fenced was that it would not remain commercially practicable or mechanically possible, that did not affect the obligation: the statute would in effect prohibit its use …'

The court pointed out that the obligation imposed by reg 4(1) is couched in unqualified terms. That meant its terms are such as to exclude considerations that would be legitimate where the obligation is qualified by referencc to thc practicability or reasonable practicability of compliance.

Use of the expression 'Every employer shall ensure …' (the opening words of reg 4(1)) will normally be indicative of the intention to impose an absolute duty.

The question in this case was whether reg 4 admitted a defence based on the cost of alternative work equipment. In answering that question, the court held that following considerations seem to be relevant.

(1) The purpose of the Directive and of the Regulation is to improve the protection of the safety and health of workers at work. Words should not be imported by implication whose effect would be to reduce the protection of the workman (Viscount Simonds in *Summers v Frost,* cited above).

(2) In case of impossibility, the Directive requires the employer to *minimise* the risk.

(3) The Directive offers no defence based on cost.

---

[9] *Edwards v NCB* [1949] 1 KB 704, per Asquith LJ at 712, endorsed by Lord Reid in *Marshall v Gotham* [1954] AC 360 at p 373.

[10] [1955] AC 740.

(4) Elsewhere in the Regulations the UK legislator has used words ('reasonably practicable') that permit a defence based on cost, but not in reg 4 or in other regulations such as reg 11 where the test is practicability alone.

(5) Where use of the words 'reasonably practicable' opens the way to a defence based on cost, that defence is strictly limited by the requirement to demonstrate gross disproportion between cost and risk. The broad construction of the word 'suitable' contended for by the defenders would be subject to no limitation at all except in so far as the courts might in the future be constrained to impose one.

(6) The broad construction contended for by the employer have would rendered the obligation imposed by reg 4 so imprecise as to be difficult to enforce and no higher than that already imposed by the common law.

(7) Adoption of such a loose construction might well bring the standard of protection afforded below that required by the Directive and so render the regulation incompatible with the Directive.

For all of these reasons they held that it is not open to the courts to adopt the broad construction of the word 'suitable' in reg 4 contended for by the defenders or to import words that would allow the defenders to advance a defence based on cost.

Compare that outcome with the quote from Swanwick J in *Stokes v Guest Keen and Nettlefold Bolts & Nuts Ltd*,[11] referred to above, that in balancing the need to deal with a 'risk' the employer must:

> 'weigh up the risk in terms of the likelihood of injury occurring and the potential consequences if it does; and he must balance against this the probable effectiveness of the precautions that can be taken to meet it and the expense and inconvenience they involve.'

Modern statute creates a completely different situation. The burden is totally different from that in negligence in many cases. For example sometimes cost can be irrelevant as *Skinner* shows.

Today we are dealing with an entirely different regime. That can be seen from the requirement for purposive interpretation discussed above. However in reality a different mindset is required.

In *Love v North Lanarkshire Council*[12] Lord McEwan showed the appropriate mindset when he considered liability under statutory duty based upon a European-based health and safety law. He commented:

---

[11] [1968] 1 WLR 1776.

[12] [2007] CSOH 10.

'It is now beyond argument that these Regulations founded on must be given a liberal and wide interpretation in order to fully implement the Directive. That that is so, finds expression in many authorities beginning with *Litster v Forth Dry Dock Etc Ltd* 1989 (HL) 96 and continuing with many other cases.

Also many of the Regulations are couched in strict language and few admit of any statutory defence. They have here variously been described as "strict" or "absolute" duties. I do not think the distinction now matters in the new climate of European Directives and increased safety. The older jurisprudence making distinctions between strict and absolute is no longer relevant.

English v North Lanarkshire Council [1999] SCLR 310 is now the accepted authority in Scotland for the need to give a wide interpretation to the Regulations, properly to implement the Directive. Lord Reed after proof followed earlier European and House of Lords authority (*Litster*) for example.

The only qualification to that approach is seen in *McGhee v Strathclyde Fire Brigade* 2002 SLT 680 where Lord Hamilton after proof said that whatever the Regulation said, foreseeability was important and the circumstances had to disclose some real risk of injury. Although the court heard argument as to whether some of the Regulations were "absolute" duties, no concluded view was expressed.'

The modern approach is again seen from Lord Reed's view in *Gallacher v Kleinwort Benson (Trs) Ltd*[13] that it is necessary to take account of the European law dimension in the interpretation and application of such regulations; and that it should not be assumed that any approach adopted under the Factories Acts continues to apply.

Lord Hamilton expressed a similar view in *McGhee v Strathclyde Fire Brigade* (para 9), the point being that the implementation of the Workplace Directive emphasises that the whole climate of health and safety has changed significantly in recent years.

He recognised that earlier case law on employers' liability must now be considered with this in mind and be viewed with great care,

'lest one blindly applies the hoops, hurdles and other stumbling blocks placed in front of a pursuer with a genuine claim by Parliament and the Courts in a less enlightened era when the health and safety of employees were regarded as less important than they are today.'

[13] [2003] SCLR 384, para 43.

3.4 THE EMPLOYER'S DUTIES

There are five basic duties that are common to *all* European health and safety regulations. The first employer's duty is to carry out a risk assessment.

3.4.1 Risk assessment

Risk assessment provisions are set out in the Management of Health and Safety at Work Regulations 1992 and 1999. They require employers and self-employed people to assess the risks created by their undertaking so as to identify the measures they need to have in place to comply with their duties under health and safety law. As such, the assessment provisions of the Management Regulations are superimposed over all other workplace health and safety legislation, including the general duties in the Health and Safety at Work etc Act 1974.

3.4.1.1 Management of Health and Safety Regulations 1992[14]

The original core regulations came in on 1 January 1993. They introduce the concept of and a requirement to carry out *risk assessments.* This is very important because it shifts the way we look at an accident to consider foreseeability. Employers must be proactive not reactive.

The regulations require employers to have arrangements in place to cover health and safety that should be integrated with management systems:

- planning
- organisation
- control
- monitoring and review.

The various HSE produced approved codes of practice and guidance on regulations provide a great deal of extra detail.

3.4.1.2 The Management Regulations 1992 in Outline

The Management Regulations 1992:

(1) required health surveillance for employees ('as appropriate');

(2) introduced requirement to appoint 'one or more competent persons for the purposes of Health and Safety assistance';

---

[14] 1992/1999.

(3) established emergency procedures for '*serious and imminent dangers*';

(4) required that employers must provide information to employees on:
   (a) risks to health identified by assessments;
   (b) preventative and protective measures;
   (c) procedures for dealing with imminent dangers.

3.4.1.3 Key Regulation

Regulation 3 of the Management of Health and Safety at Work Regulations 1992 requires the following:

> 'Every employer and self-employed person shall undertake a suitable and sufficient assessment of the risks to health and safety of employees and persons not in his employment'

The significant findings (and conclusions) of the risk assessments must be recorded in writing if the employer employs five or more persons.

This makes the Management Regulations risk assessment provisions very wide ranging and all-embracing. They are comprehensive in coverage of places, activities and other sources of hazard. They require employers to assess all the risks in the workplace that could cause harm to employers, employees and members of the public. They also need to consider the chances of a risk-causing harm.

Employers then need to decide on the precautions they must take to prevent this happening. In doing so employers will have to take account of specifics of the law. The Management Regulations in effect require employers to examine what in their work could cause harm to people so that they can weigh up whether they have taken enough precautions or should do more to meet what the law says they must do.

To recap this is what's involved. There are three elements in risk assessment – hazard / harm / risk.[15] A hazard is anything with the potential to cause harm. Harm is damage. Risk is the chance that the hazard will cause harm. Risk may be large or small. Risk assessment is the process of identifying the hazards. This is followed by assessing the risk of harm posed by the hazards then evaluating the consequences of the risk and finally using risk reduction and control measures. Having done that the second employer's duty is to take steps to control the risks found. The third employer's duty is to use the controls they put into place. The fourth

[15] HSE has laid down the following definitions: Hazard – The inherent nature of a situation or the inherent property of a material that has the ability to cause harm. A hazardous situation is one that may give rise to personal injury. Risk – The combination of the likelihood of an injury and the severity of the injury that may occur. The employer must carry out an assessment of risk to his employees and others who might be affected by his activities. It is not only employees who are meant to be given protection.

employer's duty is to monitor and maintain the controls and the fifth and final duty is to inform, instruct and train *all* involved. This is basically all that is involved in UK health and safety law whichever regulations you consider. The same five duties apply.

3.4.1.4 Burden of Proof

Burden of proof in cases where any health and safety regulations apply is reversed. The employer must prove compliance. It is not for the employee to establish the breach. For a very long time this has been accepted as the position when straightforward statutory duty is applicable. In *Bilton v Fastnet Highlands Ltd*[16] Lord Nimmo Smith confirmed that the same applies to the regulations.

The case involved a claim for damages against her employer by a prawn sorter in a fish-processing factory in respect of injury sustained by exposure to respirable prawn protein. She had developed occupational asthma and it had been exacerbated by exposure to sodium metabisulphite and sulphur dioxide in breach of reg 7 of the Control of Substances Hazardous to Health Regulations 1988 (COSHH).

The defenders contended that the pursuer must aver what steps they should have taken if she was to prove breach of the Regulations.

In allowing the pursuer proof before answer the court held that it was clear from *Nimmo v Alexander Cowan & Sons Ltd*[17] it was enough for the purposes of a case under s 29(1) of the Factories Act 1961 for the pursuer to aver that the workplace where he worked was not kept safe and he did not have to show whether it was reasonably practicable to make and keep it safe.

COSHH 1988 reg 7(1) was confirmed by Lord Nimmo Smith to be comparable with s 29(1) of the 1961 Act. The employer was accordingly under an absolute duty to make the workplace safe subject to the defence of reasonable practicality. This approach applies to all health and safety regulations.

In *Williams v Farne Salmon & Trout Ltd*[18] the pursuer alleged that he had developed occupational asthma as a result of exposure to micro-organisms in salmon. The question raised was whether on a proper construction of the COSHH Regulations employers were only bound to comply with them to the extent that they knew or ought reasonably to foresee that a substance to which an employee was exposed was a substance hazardous to health.

---

16 1998 SLT 1323, OH.
17 1967 SC (HL) 79.
18 1998 SLT 1329.

The judge noted that the definition was:

> 'couched in factual terms which are unqualified by the existence of any state of knowledge or reasonable foreseeability ... I see no difference, for present purposes, between a substance being in fact hazardous to health and a place being unsafe, and in my opinion the 1988 Regulations impose the same kind of absolute duty as is imposed by s 29(1).
>
> A number of other provisions in the regulations reinforce me in this view ... The absolute nature of this duty is, in my view, made abundantly clear by the provisions of reg 7(1), which uses the word ' ensure' in connection with the employer's duties, subject to a limited defence of reasonable practicability in respect of the duty to prevent the exposure of his employees to substances hazardous to health.
>
> The risk assessment provisions of reg 6(1), the monitoring provisions of reg 10(1) and (3), the surveillance provisions of reg 11(1) and the information, instruction and training provisions of reg 12(1) all seem to me to presuppose the actual or potential existence of an objectively verifiable state of affairs, and to place the onus on the employer to discover this, the better to ensure compliance with his absolute duty to protect his employees from exposure to substances hazardous to health'

The burden of proof is reversed.

3.4.1.5 Risk Assessments

According to the Health and Safety Executive undertaking a suitable and sufficient risk assessment, involves:

- reviewing all the activities undertaken by the organisation;
- evaluating the relevant regulations, standards and guidance;
- deciding what the organisation must do in order to ensure adequate control of risks.

The purpose is to ensure that risks are under adequate control. A risk assessment has no merit, unless it:-

- confirms that risks are under adequate control; or
- identifies actions that are necessary to ensure adequate control; and
- specifies short term measures to protect employees.

The fact that considering the relevant regulations alone is not enough was confirmed by Court of Appeal most recently in *Ellis v Bristol City*

Council.[19] They held that a code of practice that is designed to give practical guidance to employers as to how to comply with their duties under statutory regulations can be taken as providing assistance as to the meaning it was intended those regulations should have.

In this case they held that the first instance judge should have considered the meaning and purpose of the relevant regulation, any relevant judicial authority and also the approved code of practice.

They confirmed that it is well established that official publications emanating from the relevant government department can be referred to in civil proceedings as an aid to construction. Smith LJ stated:

> 'It seems to me that a Code of Practice which is designed to give practical guidance to employers as to how to comply with their duties under statutory regulations can be taken as providing some assistance as to the meaning it was intended those regulations should have.'

3.4.1.6 Management of Health and Safety Regulations 1999

These replaced the 1992 Regulations. They came Into Force on 29 December 1999. They re-enact the 1992 Regulations and add to them. The main addition was a new reg 4 dealing with principles of prevention It states:

> 'Where an employer implements any preventive and protective measures he shall do so on the basis of the principles specified in Schedule 1 to these Regulations.'

Schedule 1 (the General Principles of Prevention) specifies the general principles of prevention set out in art 6(2) of Council Directive 89/391/EEC. They provide a hierarchy approach. These are the 'Principles of Prevention':

(a) avoiding risks;

(b) evaluating the risks which cannot be avoided;

(c) combating the risks at source;

(d) adapting the work to the individual especially as regards the design of workplaces, the choice of work equipment and the choice of working and production methods, with a view, in particular, to alleviating monotonous work and work at a predetermined work-rate and to reducing their effect on health;

(e) adapting to technical progress;

---

19 [2007] EWCA Civ 685.

(f) replacing the dangerous by the non-dangerous or the less dangerous;

(g) developing a coherent overall prevention policy which covers technology, organisation of work, working conditions, social relationships and the influence of factors relating to the working environment;

(g) giving collective protective measures priority over individual protective measures; and

(h) giving appropriate instructions to employees.

It is important to remember that they provide a hierarchy approach. An employer must start at the top and try to avoid or eliminate a risk and only move down, point by point, in descending order if proper level of protection of the worker cannot be attained without moving on. It is worth remembering that the last point is often the first used to blame an injured worker. Training, instructions, information and so on are actually meant to be a 'last resort' not a first step to achieving safety.

The Court of Appeal looked at this area in February 2008 in the case of *Allison v London Underground Ltd.*[20] The judgment of Smith LJ at paragraph 57 is telling:

> 'How is the court to approach the question of what the employer ought to have known about the risks inherent in his own operations? In my view, what he ought to have known is (or should be) closely linked with the risk assessment which he is obliged to carry out under reg 3 of the 1999 Regulations.
>
> That requires the employer to carry out a suitable and sufficient risk assessment for the purposes of identifying the measures he needs to take to comply with the requirements and prohibitions imposed upon him by or under the relevant statutory provisions. What the employer *ought* to have known will be what he *would* have known if he had carried out a suitable and sufficient risk assessment.
>
> Plainly, a suitable and sufficient risk assessment will identify those risks in respect of which the employee needs training. Such a risk assessment will provide the basis not only for the training which the employer must give but also for other aspects of his duty, such as, for example, whether the place of work is safe or whether work equipment is suitable.'

Clearly there are cases where the lack of adequate risk assessment can provide a route to establish liability.

---

[20] [2008] EWCA Civ 71.

CHAPTER 4

CONTRIBUTORY NEGLIGENCE

4.1 INTRODUCTION

Contributory negligence is a constant source of debate. Before 1945 any finding of contributory negligence defeated a claim entirely. With the introduction of s 1 of the Law Reform (Contributory Negligence) Act 1945 a claim in respect of damages could not be defeated by reason of some contributory fault of the person suffering the damage.

It is important to recognise that in employer's liability cases the likely percentage level of contributory negligence in any given case will depend not just on its facts, but more importantly on the basis of the claim. In negligence cases there is scope potentially for higher reductions than in cases based upon breach of statutory duty including breach of regulations. The situation in such cases is considered in detail below. But first we need to look at basic principles.

4.2 INNOCENT CLAIMANTS

Can someone be guilty of contributory negligence when they are simply in the wrong place at the wrong time – n other words where they are not at all to blame for the accident? The answer can be yes if they made their injuries worse by not taking reasonable precautions that might have prevented or reduced them.

The answer comes from a very old employer's liability case *Jones v Livox Quarries Ltd*[1] in a quote from Denning LJ:

> 'A person is guilty of contributory negligence if he ought reasonably to have foreseen that, if he did not act as a reasonable, prudent man, he might be hurt himself; and in his reckonings he must take into account the possibility of others being careless.
>
> If a man carelessly rides on a vehicle in a dangerous position, and subsequently there is a collision in which his injuries are made worse by

1 [1952] 2 QB 608.

reason of his position than they would otherwise have been, then his damage is partly the result of his own fault, and the damages recoverable by him fall to be reduced accordingly.'

Jones, who was employed by the defendants, was proceeding from his work to the canteen for lunch. Contrary to instructions he had obtained a lift, unknown to the driver, on one of the defendants' traxcavators (a kind of excavator). Jones was standing on the towbar at the back of the vehicle. The vehicle stopped after turning a sharp bend and as a result a dumper travelling close behind ran into the traxcavator. Jones was injured and sued the defendants, alleging negligence on the part of both drivers, but the allegation against the driver of the traxcavator was not pursued.

At first instance the judge found negligence on the part of the driver of the dumper, in failing to act as a reasonably careful driver and in particular in not keeping an adequate look-out. He also found contributory negligence on the part of the plaintiff, in that he had placed himself in a position of danger on the traxcavator, and held that the plaintiff himself was 20% responsible for the damage that had occurred to him. Both parties appealed.

The Court of Appeal held that the driver of the dumper could not have been keeping any sort of look out; it was his duty to keep a proper look out, and that duty was owed to the plaintiff even though he was on the traxcavator, therefore, the defendants' cross-appeal failed.

On the question of contributory negligence they held that it was not so much a question of whether the plaintiff's conduct was the cause of the accident, as whether it contributed to the accident on the assumption that it was something of a kind that a reasonably careful man so placed would not have done. If he unreasonably or improperly exposed himself to this particular risk, he ought not to be allowed to say that it was not a cause operating to produce the damage, even though the prohibition against riding on the vehicle had not been made with that particular risk in mind. The judge had not been wrong in finding that the plaintiff, who deliberately put himself into a position which exposed him to this danger, was to some extent responsible for what had happened, and, therefore, the plaintiff's appeal was also dismissed.

Jones v Livox Quarries Ltd has formed the basis of many later key decisions particularly, *Froom v Butcher*.[2] The plaintiff, who was not wearing the fitted seat belt while driving his car, received head and chest injuries and a broken finger as the result of a collision solely attributable to the negligent driving of the defendant. Except for the broken finger, the injuries would probably have been prevented by the wearing of a seat belt.

2 [1976] QB 286.

The plaintiff stated in evidence that he did not like wearing the belt because of the danger of being trapped in the vehicle after the crash and because he did not drive at more than normal speed. Nield J held that the omission to wear the seat belt did not amount to contributory negligence upon the basis that to find otherwise would constitute an unjustified invasion of the freedom of choice of the motorist; if he was wrong, he would have reduced the damages by 20%.

The Court of Appeal, allowing the defendant's appeal, held that a reduction of 20% should be made just as it was in *Jones v Livox Quarries Ltd.* They held that the prudent man should guard against the possibility of negligence by others by wearing a seat belt; that the chances of injury are four times as great when a seat belt is not worn; that in determining whether contributory negligence is made out the proper question is not what or who caused the accident but what was the cause of the damage.

Lord Denning gave the leading judgment in this case and indicated that a person who failed to wear a seat-belt must share some responsibility for the damages. If the evidence showed that the injuries would have been prevented altogether if a seat-belt would have been worn then the damages would be reduced by 25%, if the injuries would have been less severe if a seat-belt had been worn then the reduction would be 15%. Only if the failure to wear a seat-belt made no difference at all to the injuries sustained would the injured party not suffer any deduction in his or her damages claim.

That is still the position and arguments that such deductions should be increased as the wearing of seatbelts was not compulsory then are irrelevant. It followed the previous law. It was intended that the wearing of seatbelts was to be made compulsory and there had been a massive public awareness campaign to get people to wear them.

Jones v Livox Quarries Ltd even dictates the approach to deductions for contributory negligence for a passenger injured in a road traffic accident when he knows the driver has consumed enough alcohol to impair his ability to drive safely, or if he goes drinking with the driver knowing he will be a passenger later when the drink deprives him of his own capacity to appreciate the danger. The case here is *Owens v Brimmell.*[3]

Owens and Brimell went out in Brimell's car for a pub-crawl. They both drank a great deal. On the way home there was an accident, caused by Brimell's negligence, and Owens suffered severe and permanent injury. Later Brimell pleaded guilty to driving without due care and attention and with excess alcohol in his blood. Owens claimed damages. Brimell contended that the damages should be reduced due to the claimant's contributory negligence.

---

[3] [1977] 1 QB 859.

If the burden of proving contributory negligence is discharged then the question is what the level of the deduction should be in a case of this type. Watkins J gave the clearest exposition of the principles of law for determining whether a passenger was contributory negligent in riding with an inebriated driver.

In defining contributory negligence Watkins J quoted Lord Denning MR's dictum in *Froom v Butcher*:

> 'Contributory negligence is a man's carelessness in looking after his own safety. He is guilty of contributory negligence if he ought to have foreseen that if he did not act as a reasonably prudent man he might be hurt himself.'

The judge held that Owens had been guilty of contributory negligence, but only to the extent of 20% as in *Jones v Livox Quarries Ltd.* A deduction of 20% is still seen as the top limit in this type of case.

The impact of *Jones v Livox Quarries Ltd* continues to be seen frequently. In *Badger v Ministry of Defence*[4] Beryl Badger made a claim for damages following the death of her husband through exposure to asbestos. The damages were reduced by 20%, as her husband had contributed to his own death by refusing to give up smoking, which also caused the lung cancer that killed him.

Stanley Brunton J held that a reasonably prudent man, warned that there was a substantial risk that smoking would seriously damage his health, would stop smoking. *Jones v Livox Quarries Ltd* was the starting point for the consideration of a deduction in this case too.

4.3 100% CONTRIBUTORY NEGLIGENCE?

The word 'negligence' in the expression 'contributory negligence' sometimes seems to distort the view of judges. The following passage in *Prosser & Keeton on Torts*[5] appears as a footnote in *Hickey v Zezulka*[6] and is very much to the point:

> 'It is perhaps unfortunate that contributory negligence is called negligence at all. "Contributory fault" would be a more descriptive term. Negligence as it is commonly understood is conduct which creates an undue risk of harm to others. Contributory negligence is conduct which involves an undue risk of harm to the actor himself. Negligence requires a duty, an obligation of conduct to another person. Contributory negligence involves no duty, unless we are to be so ingenuous as to say that the plaintiff is under an obligation to protect the defendant against liability for the consequences of the plaintiff's own negligence.'

---

4 [2005] EWHC 2941, QB.

5 (5th edn, 1984) s 65, p 453.

6 (1992) 487 NW 2d 106.

This concept of 100% contributory negligence being unsustainable is true in all actions. A good example is *Pitts v Hunt*,[7] a famous road traffic accident based on negligence. When considering whether the appellant should have his damages reduced to nil because of his own contributory negligence Beldam LJ concluded **rightly** that the wording of s 1 of the Law Reform (Contributory Negligence) Act 1945 is incapable of a such an interpretation.

> 'Section 1 begins with the premise that the person suffers damage as a result partly of his own fault and partly of the fault of any other person or persons. Thus before the section comes into operation, the court must be satisfied that there is fault on the part of both parties which has caused damage.
>
> It is then expressly provided that the claim shall not be defeated by reason of the fault of the person suffering the damage. To hold that he is himself entirely responsible for the damage effectively defeats his claim.
>
> It is then provided that the damages recoverable in respect thereof (that is the damage suffered partly as a result of his own fault and partly the fault of any other person) shall be reduced. It therefore presupposes that the person suffering the damage will recover some damage.
>
> Finally reduction is to be to such extent as the court thinks just and equitable, having regard to the claimant's share in the responsibility for the damage. To hold that the claimant is 100% responsible is not to hold that he shared in the responsibility for the damage.'

This view was echoed by Balcombe LJ:

> 'I agree that the judge's finding that the plaintiff was 100% contributorily negligent is logically unsupportable and, to use his own words, "defies common sense".'

A finding of 100% contributory negligence would be equivalent to saying that the claimant was solely responsible for his own injuries. The correct position was summed up by Sedley LJ in *Anderson v Newham College of Further Education*[8] when he expressed the view that *Jayes v IMI (Kynoch) Ltd*[9] should not be followed by judges of first instance and should not be relied upon by advocates in argument.

---

[7] [1990] EWCA Civ 17.

[8] 2002 EWCA Civ 500.

[9] (1985) ICR 155. There it was wrongly held that: '... there is no principle of law which requires that, even where there is a breach of statutory duty in circumstances such as the present, (where the intention of the statute is to provide protection, *inter alia*, against folly on the part of a workman), there cannot be a case where the folly is of such a kind or of such a degree that there cannot be 100% contributory negligence on the part of the workman.'

> 'The relevant principles are straightforward. Whether the claim is in negligence or for breach of statutory duty, if the evidence, once it has been appraised as the law requires, shows the entire fault to lie with the claimant there is no liability on the defendant. If not, then the court will consider to what extent, if any, the claimant's share in the responsibility for the damage makes it just and equitable to reduce his damages. The phrase "100% contributory negligence", while expressive, is unhelpful, because it invites the court to treat a statutory qualification of the measure of damages as if it were a secondary or surrogate approach to liability, which it is not. If there is liability, contributory negligence can reduce its monetary quantification, but it cannot legally or logically nullify it.'

So the conclusion is that even in cases simply based upon negligence, if the employer's negligence caused the accident the claimant will always recover some compensation. As will be seen in the next section this also applies in cases based upon breach of statutory duty but with much more impact in terms of limiting reductions.

4.4 BREACH OF STATUTORY DUTY AND CONTRIBUTORY NEGLIGENCE

As we have already seen a finding of 100% contributory negligence would be equivalent to saying that the claimant was solely responsible for his own injuries. When there is a causative breach of statutory duty then clearly that is not the case. Goddard LJ's remarks in *Hutchinson v London and North Eastern Railway Co*[10] form a good starting point for some thoughts on this decision:

> 'It is only too common to find in cases where the plaintiff alleges that a defendant employer has been guilty of a breach of a statutory duty, that a plea of contributory negligence has been set up. In such a case I always directed myself to be exceedingly chary of finding contributory negligence where the contributory negligence alleged was the very thing which the statutory duty of the employer was designed to prevent.'

Another very important case to remember is *Westwood v Post Office*.[11] Westwood was employed by the Post Office at a telephone exchange as a technical officer. Employees at the exchange were in the habit of going on to its flat roof for a 'breather'. That practice was known to and not objected to by the Post Office. The normal means of access to the roof was through one of two doors at the tops of stairways at each side of the roof. On the day in question Westwood with other employees mounted one of the stairways, intending to go on to the roof, and found the door to the roof locked but an adjacent door into the lift motor room ajar. It was possible to get on to the roof through the motor room window. That means of access was used by employees from time to time, but the practice was not known to the Post Office or their responsible officers. On the

10 [1942] 1 KB 481, 488.
11 [1973] 3 WLR 287.

door of the motor room was a notice reading: 'Notice. Only the authorised attendant is permitted to enter'.

Westwood and his companions used the route through the motor room to get to the roof, and when their break was over returned the same way. On the way back Westwood stepped on a trap door in the floor of the motor room; it gave way under his weight and he fell through to the floor beneath and was fatally injured. The trap door was of inadequate construction and insufficient strength as a floor.

The case was based on breach by the Post Office of their statutory duty under s 16 of the Offices Shops and Railway Premises Act 1963. O'Connor J held that the Act applied to the lift motor room and gave judgment for the plaintiffs, rejecting an allegation of contributory negligence against Westwood by the Post Office.

The Lords held that the fact that Westwood had gone there for his own purposes unconnected with his duties did not deprive him of the protection of the Act. The wording of the notice had not suggested that there was any danger in the room and the fact that a man was a trespasser had no bearing upon whether he might reasonably foresee harm to himself.

Westwood was described as having been disobedient but not negligent, and so in relation to the Post Office's breach of statutory duty had not been guilty of any contributory negligence. Lord Kilbrandon said:

> 'My Lords, the defence of contributory negligence as an answer, even as nowadays only a partial answer, to a claim arising out of breach of statutory duty is one which it must always be difficult to establish. The very existence of statutory safety provisions must be relevant to the consequences which a man may reasonably be expected to foresee as arising from his own conduct; his foresight as to that will be to some extent governed by what he may reasonably be expected to foresee as arising from his master's statutory obligations.'

He confirmed the decision in *Grant v Sun Shipping Co Ltd*[12] quoting Lord du Parcq at p 567:

> 'I am far from saying that everyone is entitled to assume, in all circumstances, that other persons will be careful. On the contrary, a prudent man will guard against the possible negligence of others when experience shows such negligence to be common. Where, however, the negligence is a breach of regulations, made to secure the safety of workmen, which may be presumed to be strictly enforced in the ordinary course of a ship's discipline, I am not prepared to say that a workman is careless if he assumes that there has been compliance with the law. The real complaint of the defenders is that the pursuer reposed an unjustified confidence in them. No doubt his

---

12 [1948] AC 549.

> confidence was not justified in the event, but he is not, I think, to be blamed for that. The courts have long recognised that in some circumstances an omission to make sure for oneself that others have done what they ought to have done is not negligent.'

Clearly this case means that in many cases there can be no discount at all. If there is then usually there is no basis for a discount above a maximum of 50%. This is on the authority of the House of Lords in *Boyle v Kodak Ltd*[13] recently confirmed by the Court of Appeal in *Anderson v Newham College Of Further Education*.[14]

In *Boyle* the appellant, who had been employed to paint the top of an oil storage tank at the respondent's works, climbed a ladder, instead of an external iron staircase, in order to lash it at the top of the tank. In the attempt he fell and was injured. There was a breach of the Building (Safety, Health and Welfare) Regulations 1948, reg 29(4) the relevant part of which read:

> 'Every ladder shall so far as practicable be securely fixed so that it can neither move from its top nor from its bottom points of rest.'

The employer had never instructed the appellant to use the iron staircase which gave access to the top of the tank for this purpose, but assumed that, being a very experienced workman, he would comply with the statutory requirement. The trial judge found the appellant solely to blame for the accident and the Court of Appeal upheld his decision.

The House of Lords held that on a claim for damages for breach of statutory duty, an employee need do no more than to establish that the employer was in breach of an absolute statutory duty and, unless the employer can show that he did take all reasonable steps to prevent the breach,[15] he may not escape liability even where negligence at common law cannot be established against him. Their lordships held that the employers had not proved that they did all that could be reasonably expected of them for ensuring compliance with the regulation. Equal apportionment was the outcome.

It is important to remember what Lord Tucker said in *Staveley Iron & Chemical Co Ltd v Jones*:[16]

> '[I]n Factory Act cases the purpose of imposing the absolute obligation is to protect the workmen against those very acts of inattention which are sometimes relied upon as constituting contributory negligence so that too strict a standard would defeat the object of the statute.'

---

13 [1969] 1 WLR 661.

14 [2002] EWCA Civ 505.

15 As with reg 12(3) of the Workplace (Health Safety and Welfare) Regulations 1992, SI 1992/3004.

16 [1956] AC 627.

In *Reeves v Metropolitan Police Comr*[17] Lord Hoffmann pointed out that the question to be determined is the relative responsibility of the two parties, not degrees of carelessness. That question has to take into account the policy behind the rule by which the liability is imposed. Regulations are designed, at least in part, to protect the employee from the consequences of his own negligence.

Referring to what Lord Tucker said in *Staveley Iron & Chemical Co Ltd v Jones* Lord Hoffmann said:

> 'This citation performs the valuable function of reminding us that what section 1 requires the court to apportion is not merely degrees of carelessness but "responsibility" and that an assessment of responsibility must take into account the policy of the rule, such as the Factories Acts, by which liability is imposed. A person may be responsible although he has not been careless at all, as in the case of breach of an absolute statutory duty. And he may have been careless without being responsible, as in the case of 'acts of inattention' by workmen.'

In other words 'responsibility' for the breach lies with the employer whether the employee has been careless or not, as in the case of breach of an absolute statutory duty. What employers often call contributory negligence is often no more than inattention or inadvertence and does not attract any deduction.

4.5 TECHNICALITIES

It is important to remember that contributory negligence must be pleaded and proved by the defendant. A finding of contributory negligence cannot be based on an allegation that is neither pleaded nor argued at trial.[18] In addition no reduction should be made unless, on a balance of probabilities basis, the defendant reaches the required standard. That standard is often hard to achieve.

In *Dawes v Aldis & NIG PLC*[19] Eady J reminded us of just how hard it can be for a defendant to obtain a finding that on the balance of probabilities there must have been contributory negligence on the part of a claimant. He concluded that he could not reach a decision on the balance of probabilities, so as to be able to draw the inference that there must have been contributory negligence, saying: 'It remains, of course, a strong possibility but I cannot elevate it to a probability'.

So a strong probability is not enough to give rise to a deduction. A similar approach was taken by His Honour Judge Mackie sitting as a High Court

---

[17] [2000] 1 AC 360 at p 371.

[18] For a recent case confirming this see *Dziennik V CTO Gesellschaft Fur Containertransport MBH & Co MS Juturna KG* [2006] EWCA Civ 1456.

[19] [2007] EWHC 1831, QB.

judge in *Wakling v McDonagh*[20] when he referred to the fact that the test for contributory negligence as set out in the Law Reform (Contributory Negligence) Act 1945, s 1(1) requires that the defendant must establish on the balance of probabilities that:

(i) the claimant was at fault;

(ii) the fault was causative of the injury suffered: and

(iii) it would be just and equitable for his damages to be reduced.

The third point is too often ignored. Conduct by a defendant could mean that on the third test alone there should be no reduction.

What of the claimant who is careless or not paying attention to what he was doing? The first thing to consider is whether that is really just inattention or inadvertence. As can be seen from the earlier cases there is ample support for the proposition that inattention or inadvertence does not form the basis for a finding of contributory negligence on the part of an employee who has been injured as a result of a breach by his employers of an absolute statutory duty. This was confirmed yet again by Lord Nimmo Smith in *McGowan v W & J R Watson Ltd*[21] when he said:

> 'The reason for this is that statutory provisions of this kind are intended to protect employees against *inter alia* accidents caused by inattention or inadvertence. The protection does not extend only to employees who are fully alert. A momentary lapse, such as occurred in the present case, falls short of being described as a lack of reasonable care on the part of the pursuer.'

Even when a claimant is well trained, very experienced, quite senior and ignores safety rules the best that a defendant can often hope for is a deduction for contributory negligence. In *Ashbridge v Christian Salvesen plc*[22] Lord Glennie was concerned with regulations designed to protect the employee against dangers inherent in working with machinery and equipment of various kinds. He accepted that it is a feature of every working environment that there will be moments of carelessness or lack of concentration saying:

> 'It is in part to guard against danger arising in such an environment from such carelessness or lack of concentration that the Regulations assume a role of great importance. It follows that the purpose of the Regulations would be defeated if a finding of contributory negligence were made whenever an employee was careless and by his carelessness contributed to the accident. This applies *a fortiori* to careless or sloppy practices which have become rife and of which the employers are, or ought to be, aware. It is,

---

20 [2007] EWHC 1201, QB.

21 [2006] CSIH 62.

22 [2006] CSOH 79.

> therefore, the exceptional case rather than the norm where a finding of contributory negligence will be made.'

However on the facts he held that in this case Ashbridge's actions went beyond the sort of carelessness or inadvertence that he described concluding: 'It seems to me that the pursuer was guilty of the most wanton disregard of his own safety'. Even so the judge recognised that this did not absolve the defenders from responsibility for the inadequacies in their system. He assessed Ashbridge's responsibility for the accident at 50%.

PART B

STATUTORY FRAMEWORK

CHAPTER 5

ACCIDENTS IN THE WORKPLACE

5.1 INTRODUCTION

The Workplace (Health, Safety and Welfare) Regulations 1992[1](the 1992 Regulations) that came into force, (subject to certain exceptions and transitional provisions), on 1 January 1993 sought to provide a uniform set of duties to all workplaces. In doing so they largely replaced the duties under the Factories Act 1961 and the Office Shops and Railway Premises Act 1963. The 1992 Regulations implemented the European Workplace Directive 1989.

Generally speaking the duties in the Regulations are wider and stricter than the common law duties established over time. Indeed, in dealing with workplace accident claims the principal focus of the investigation will be on the Regulations rather than the common law principles of negligence.

Transitional provisions applied; workplaces in existence as at 1 January 1993 were not subject to all of the Regulations until 1 January 1996. There were special provisions in relation to modifications, extensions or conversions that were not completed up until 17 September 2002; but from that date all modifications, extensions or conversions became subject to the Regulations regardless of when started.

In addition there is an approved code of practice (ACOP) and further information on guidance on how the Regulations are to facilitate safety at work can be found in various publications, specific to various industries, published by the Health and Safety Executive (HSE) and there is considerable information available on the HSE website. These are valuable resources and need to be considered when dealing with claims.

The Health and Safety at Work etc Act 1974 (the HSWA 1974) makes the ACOP admissible in criminal proceedings but is, oddly, silent in respect of civil proceedings. However, it is common place for the codes and guidance to be cited and relied upon in civil proceedings. This goes hand in hand with the semi exclusion of expert evidence in the post-Woolf era but

[1] SI 1992/3004.

caused a difficulty in *Ellis v Bristol City Council*[2] where the Court of Appeal overturned the first instance judge who had declined to take into account submissions based upon the ACOP. In that case the ACOP was not raised in cross-examination and the Court of Appeal confirmed that was unnecessary. However, it can be used in that way if the particular case requires it to illustrate best practices.

5.2 WHAT CONSTITUTES A WORKPLACE

The definition is found at reg 2 of the 1992 Regulations. Workplace means, subject to para 2, any premises or part of premises which are not domestic premises are made available to any person as a place of work and includes:

> any place within the premises to which such person has access while at work; and
>
> any room, lobby, corridor, staircase, road or other place used as a means of access to or egress form that place of work or where facilities are provided for use in connection with the place of work other than a public road.

Paragraph 2(2) dealt with the situation re extensions and modifications but, as of 17 September 2002, is no longer relevant.[3]

Essentially, the workplace is an area designated for people to work in and includes the means of getting to and from it. In the vast majority of cases the concept will not give rise to any difficulty and the situation will be taken as read.

Not all places at which work is carried out are covered and there are exemptions in relation to ships, most but not all construction sites[4] and mines. In addition aircraft, locomotive or rolling stock, trailers or semi-trailers used as a means of transport or a vehicle by licence under the Vehicles (Exercise) Act 1971. The reason for the exclusions is that ships in dock, mines and construction places have their own special provisions; and in respect of aircraft and other vehicles, it will readily be understood that it would not be possible to apply the strict regulations to vehicles on the road or to airplanes in flight.

In relation to vehicles etc there is a limited, but noteworthy, exemption to the exclusion of the Regulations and that is when an aircraft or vehicle is

---

2 [2007] EWCA Civ 685.

3 SI 1992/3004, reg 2(2).

4 From 6 April 2007 the words in reg 3(1)(b) were substituted by 'a workplace which is a construction site within the meaning of the Construction (Design and Management) Regulations 2007 and in which the only activity being undertaken is construction work within the meaning of those regulations, save that— (i) regulations 18 and 25A apply to such a workplace; and (ii) regulations 7(1A), 12, 14, 15, 16, 18, 19 and 26(1) apply to such a workplace which is indoors'.

stationary in a workplace. In such circumstances the provisions relating to falls or falling objects (see below) are applicable.

5.3 WORK

HSWA 1974, ss 52 and 53 deal with the definitions of the expressions 'work' and 'at work'. In most cases it will be obvious whether someone is working or not and indeed whether they are at work. However, specific legislation does provide an extension or clarification of the situation so that work experience on certain training schemes, training which includes operations involving ionising radiations and activities involving genetic manipulation are all designated as being at work.

In *Robb v Salamis*[5] the Inner House held that a person is at work for the whole of the time when he was in the course of his employment that includes rest periods and when travelling between periods of work. This point was not the subject of the subsequent appeal to the House of Lords.

5.4 DUTY HOLDERS

Regulation 4 of the 1992 Regulations provides that the person having control of the workplace modification, extension or conversion is the person who has the responsibility for complying with the provisions of the Regulations. It states:

- an employer who controls the workplace where any of its employees work;
- any person who has, to any extent, control of the workplace; and
- any person who is deemed to occupy factory premises by virtue of the Factories Act 1961, s 175(5).

This provision deals with attempts to evade responsibility of the statutory provisions. It deems a person as an occupier of premises where factory work is carried on by two or more persons, with the permission or agreement of the occupier, to be an occupier of a factory and imposes the statutory duties on that person as if he was an employer. Factory work is essentially the making, repairing, finishing or preparing for sale any article, and includes slaughter houses.[6]

It can be seen that the primary duties is placed upon an employer but reg 4(2) extends the responsibility under the regulation to any person who has to any extent control of the workplace

5 [2006] UKHL 56.

6 Section 175(1)

This brings into play a common situation where there is multi-occupancy of a building, particularly, office buildings A difficulty may arise where a slippery substance is found to be the cause of the accident on stairs or a corridor that is in the control of the landlord or another occupant. Such cases require investigation of the surrounding circumstances to establish the responsibilities of the potential parties. This provision imposes the stricter provisions of the regulations upon the landlord of such premises but only in so far as compliance relates to matters under his control. Typically this will mean that the landlord will be responsible for the suitable construction and maintenance of the stairs and the general fabric of the building.

Control is a question of fact that has been considered in the chapter on the Construction Regulations.[7] Specifically, in *Bailey v Command Security Services Ltd*[8] a security guard fell down an unguarded lift shaft while patrolling a warehouse. The security guard's employer was held to be an occupier and had power to report such dangers but that did not establish control because there were no steps the employer could have taken to rectify the position. The proper defendant in such a case was the warehouse owner who had the power to alter the workplace.

The situation may however be different where the employer controls the way in which the work is done. That was the situation in *King v RCO Support Services Ltd*[9] In that case the claimant slipped on the yard of a bus company but he was employed by a contractor who had been brought in to grit the icy yard. The ice and the method of removing the problem were not within the bus company's control but the employer had control over how and when the work was carried out.

There can also be situations when more than one person has control: see *McCully v Farenden*.[10]

5.5 WHO IS PROTECTED?

The person who has control of the premises used as a workplace owes the duty to ensure that the regulatory scheme is complied with. Therefore, it must follow that any worker in the workplace regardless of whether he is employed by the person in control can bring an action in respect of any breach of the regulatory regime.

However, difficulties have arisen in respect of people meeting with accidents in shops and other parts of premises.

---

7 See Chapter 10.
8 (unreported) 25 October 2001, QB.
9 [2001] ICR 608.
10 2003 NIQB 6.

A number of Scottish decisions have indicated that the meaning of the word 'person' in reg 12(3) is not confined to persons working in the shop. One other Scottish case, before Sherriff Ross, took a different view saying that a 'person' in reg 12(3) was a person at work in the workplace. The only authority in the English Court of Appeal is in the case of *Ricketts v Torbay Council*.[11] There, on an application for leave to appeal the claimant's application was dismissed; as such it cannot be cited.[12] However, it so well known that a discussion of the judgment is warranted here. Mrs Ricketts tripped as a result of a defect in a car park used by council employees and brought a claim. It was argued that the provisions applied to any person – but the real adventure was to try and argue that the provisions of reg 5 imposed an absolute duty to maintain the surface of the car park in a safe condition. The argument went that the hole was a breach and therefore if the regulations applied liability would be established. (See below on the question of the strictness of the duty.) In short the Court of Appeal held that the recorder had considered the matter and that the duty was not absolute.

Higher Courts in Scotland have produced two decisions. Firstly, in *McCondichie v Mains Medical Centre*[13] Lord Drummond Young held that the Regulations did not apply in such circumstances. Secondly, and more authoritatively, the Inner House in *Donaldson v Hays Distribution Services*[14] held that the Regulations did not extend to protect non-workers present in the workplace. The reasoning for the decision was as follows:-

(1) the fact that the Workplace Regulations were enacted to give effect in the United Kingdom to the Workplace Directive, which applies solely for the protection of worker;

(2) the absence of any positive indication in the legislation that it was intended that the Workplace Regulations should extend protection to those coming onto premises as visitors and not workers;

(3) the extreme improbability that the legislative intention was to supersede much of the law of occupiers liability without an express repeal or amendment of the Act.

The last point is probably the most telling aspect of the judgment. Although the decision is not binding on English courts it is clearly of considerable importance and may be followed. In *Ricketts* the Court of Appeal used similar reasoning to come to the same conclusion. It should also be noted that the definition of workplace applies to those at work and it is difficult to see how that would have applied to Mrs Ricketts. Even so there is no reason why a breach of the regulations cannot be

[11] [2003] EWCA Civ 613.
[12] See Practice Direction (Citation of Authorities) [2001] 1 WLR 1001.
[13] [2003] Scot CS 270.
[14] [2005] CSIH 48.

relied upon as proof a breach of duty giving rise to a finding of negligence. An example of this approach can be found in *Poppleton v The Trustees of the Portsmouth Youth Activities Committee*.[15]

In the light of *Ward v Tesco*[16] where Lord Denning placed the burden of proof on the supermarket operator to prove that all reasonable care had been taken the debate is somewhat academic – unless the duty is absolute. It becomes more so because in such cases the invariably well-prepared defendant comes to court armed with evidence to establish that everything possible was done to prevent the accident.[17] If there is a difference between the common law test in *Ward* and the test of reasonable practicality in the regulations it is a thin one; it could be said that it is an exercise in splitting hairs when there is no hair to split.

However, the situation will be different if the worker attends the shop in the course of working for another employer: see *King v RCO* above. Firstly, persons delivering goods to the shop will be subject to conditions controlled by the occupier at the shop. Further, it is not uncommon for sales representatives to work on the floor of a supermarket offering double glazing, credit cards or demonstrating goods on sale in the shop. They are clearly at work and there is no reason why the regulations should not apply – if they are not employed by the shop keeper/occupier. Another common situation is where a worker goes to the shop to buy groceries or stationery supplies. It cannot be doubted that the individual is then working and the shop owner has control over the premises. Therefore, it is submitted that the regulations do apply just as much as if the worker was in the office – but that the duty, under the regulations is on the occupier.

5.6 THE SPECIFIC DUTIES

The two most important regulations that will feature prominently in litigation arise under regs 5 and 12 and the chapter will deal with these in detail and consecutively.

> **5. Maintenance of workplace, and of equipment, devices and systems**
>
> (1) The workplace and the equipment, devices and systems to which this regulation applies shall be maintained (including cleaned as appropriate) in an efficient state, in efficient working order and in good repair.
> (2) Where appropriate, the equipment, devices and systems to which this regulation applies shall be subject to a suitable system of maintenance.
> (3) The equipment, devices and systems to which this regulation applies are—

[15] [2007] EWHC 1567, QB.
[16] [1976] 1 All ER 219.
[17] The staff state that they keep an eye out for spillages, that there is a janitor on call while the shop is open and that they clean as they go.

(a) equipment and devices a fault in which is liable to result in a failure to comply with any of these Regulations; and
(b) mechanical ventilation systems provided pursuant to regulation 6 (whether or not they include equipment or devices within sub-paragraph (a) of this paragraph).

It should be noted first of all that the duty is absolute and the use of the words 'shall be maintained' is an imperative instruction to the employer under the Regulations. The regulation is therefore a potent line of attack in a personal injury claim.

Thus in *Malcolm v Commissioner of Police of the Metropolis*[18] His Honour Judge McDuff QC (sitting in the High Court) gave judgment to the claimant in a case where her left arm was trapped when the defective doors of a lift closed on her. It mattered not how good the maintenance system was or that the defect was unforeseeable – there was strict liability. The judge noted that the wording of the regulation was derived from the definition of 'properly maintained' in the Factories Act 1961 and that in *Galashiels Gas Co Ltd v O'Donnell (or Millar)*,[19] the Privy Council had held that the duty under the Act was an absolute one. Regulation 5 was not limited by any 'reasonable practicability' clause therefore the duty owed was absolute in the same way as a duty under the 1961 Act.

The usefulness of the regulation has been curtailed in some cases but given the absolute nature of the duty it is not surprising that efforts have been made to extend its application where the reasonably practicable duty under reg 12 would not necessarily have seen judgment for the claimant. However this case related to a flight of steps and reg 12(5) only relates to staircases.

In *Coates v Jaguar Cars*[20] the Court of Appeal said that a failure to provide a handrail was not a failure to maintain but was really an allegation that something should have been provided. (In this respect see reg 12(5) which contains the duties in respect of handrails.)

In *Ball v Street*,[21] it was said that something may work or serve a function but the question of whether it is in efficient working order and in good repair raises the question of whether it is safe. This is a question of fact and degree to be considered by the trial judge.

The word 'maintain' was used in the former Factories Act 1961, s 25(1), and was considered by the House of Lords in *Latimer v AEC Ltd*.[22] There a sudden and unexpected flooding of the floor created a dangerous

---

18 [1999] CLY 2880.
19 [1949] AC 275.
20 [2004] EWCA Civ 337.
21 [2005] EWCA Civ 76.
22 [1953] AC 643.

situation and the worker slipped and fell. The argument went that the floor was not maintained in a safe condition because the flood made it dangerous.

Lord Oaksey stated that a floor does not cease to be in an efficient state because there is a piece of orange peel or a pool of water on it.

What it is important to note from this decision is that Lord Oaksey did not confine the word maintenance to its construction. He said:

> 'I think the obligation to maintain them in an efficient state introduces into what is an absolute duty a question of degree as to what is efficient.'

The matter was revisited in the case of *Lewis v Avidan Ltd*[23] where a workplace floor was again flooded unexpectedly. May LJ said:

> 'Perhaps it could be said that a flood could make the floor not in an efficient state; but those words have to read it there context. The workplace, including the floor, has to be maintained, including cleaned as appropriate, in an efficient state and efficient state appears in conjunction with efficient working order and good repair. The word maintained imports the concept of doing something to the floor itself such as cleaning or repairing it. The mere fact of a flood does not mean that the floor is not maintained in an efficient state.'

May LJ went on to dismiss the submission by counsel for the claimant that even a small amount of water would produce a breach of reg 5. It is clear that May LJ, and also Lord Oxley, in considering the question of maintaining the floor in the safe condition were looking at the question of structural integrity rather than the problems arising from transient dangers.

Many commentators consider this to be the wrong approach and it conflicts with a long line of authorities from Scotland. For example in *Love v North Lanarkshire Council*[24] dealing with reg 5 Lord McEwan noted that the regulation is couched in strict terms. Love was injured by flying glass from a bottle that was smashed when a piece of wood was accidentally knocked over. Lord McEwan held on the evidence the locker room was part of Love's workplace. Bearing in mind that he room was small, was used by many people and the evidence showed that little attention was paid to its cleanliness or tidiness he held:

> 'In my view the Regulation is to secure the outcome of a continuing state of efficiency. *Butler v Grampian University Hospital NHS Trust* 2002 SLT 985 was a case where a hospital worker was injured while helping a disabled patient from a wheelchair within a toilet cubicle. Regulation 5 was founded

---

23 [2005] EWCA Civ 670.

24 [2007] CSOH 10.

on. At Procedure Roll the Lord Ordinary held that the Regulation required the employer to secure that the workplace was in a continuing state of efficiency.

McLaughlin v East and Midlothian NHS Trust 2002 SLT 387 was also decided before proof. The pursuer was injured when a curtain rail fell unexpectedly on to her. She founded on Regulation 5. The Lord Ordinary held succinctly and without any clear analysis that the duty was strict and if a workplace was not in efficient working order liability was established.'

Lord McEwan moved on to consider other cases saying:

'*Gillanders v Bell* [2005] CSOH 54 concerned Regulation 5 and shelving which could not support a leaning ladder due to protruding bobbins. The Lord Ordinary (agreeing with Butler) held that the word 'maintained' simply meant 'secured'.

To the same effect is *Gilmour v East Renfrewshire Council*[25] where a teacher slipped on a chip left on a lino covered ramp leading from a school canteen. In a carefully reasoned opinion the temporary judge held that reg 5 was in absolute terms and might overlap with other regulations. Agreeing with *Butler* he held that reg 5 required a continuing state of efficiency.[26]

Lewis v Avidan[27] concerned a nurse who slipped on a pool of water in a nursing home hallway. The cause was a flow of water from a burst concealed pipe. The Divisional Court gave a meaning to the word 'maintained' (ie doing something) that conflicts with the Scottish cases on reg 5. I find the reasoning unattractive and do not propose to follow it.

That maintenance and repair are not the test under reg 5 is also seen in *Cochrane v Gaughan*,[28] again a case about a wet floor …I do not think that maintenance of the room is a relevant issue. If this is a correct view then the presence of the wood where it was and the glass bottle on the floor render the workplace 'not efficient'.'

It is also worth considering the House of Lords decision in *Smith v Cammell Laird*.[29] There it was held that the obligation under reg 31 of the Shipbuilding Regulations 1931 meant that the occupiers of a factory had a duty to ensure that all staging should be securely constructed and maintained in such a condition as to ensure the safety of all persons. The regulation was an absolute one; an absolute obligation was imposed through the words 'all staging … shall be maintained'. *Smith* is really a

25 (2004) Rep LR 40.

26 Lord McEwan distinguished a contrary Sheriff's Court Decision of *McNaughton v Michelin Tyres* 2001 SLT (Sh Ct) 67 and stated that he proposed to say no more about it.

27 [2005] EWCA Civ 670.

28 2004 SCLR 1073.

29 [1940] AC 242.

'workplace' case rather than a 'work equipment' case. There is an argument that *Lewis* should have followed it.

The Court of Appeal also considered the situation in *Green v Yorkshire Traction Co Ltd*,[30] in respect of reg 6(1) of the Provision and Use of Work Equipment Regulations 1992. In that case there was rainwater on the step of a bus and the Court of Appeal held that while the water created a risk it was not such as to render the bus to be not in an efficient state; it was in efficient working order and in good repair.

Therefore, while the use of the word 'maintained' indicates an absolute duty, the judiciary appear willing to link efficiency with safety aspects indicating that there is leeway and the workplace need not be perfect. The question is really whether the movement from perfection makes the workplace unsafe. However, it does not mean that deterioration needs to be foreseeable. Therefore, if, for example there is a sudden collapse of part of the building the person having control will be liable because of the absolute nature of the duty.

The position would have been different in both *Latimer* and *Lewis* if the flood had been caused by a faulty valve even if it could not be said to be a foreseeable event. The failure of such equipment would be covered by reg 5(3)(a) because it would be equipment or a device which if it failed would lead to a breach of reg 12. In such a case the claimant would rely upon reg 5, establish a breach of an absolute duty and establish primary liability.

5.7 TRIPS AND SLIPS AT WORK

The most common type of case that the practitioner will encounter is the tripping or slipping situation. It is therefore worthwhile setting out the whole of the governing regulation. Regulation 12 provides as follows:

> **12. Condition of floors and traffic routes**
>
> (1) Every floor in a workplace and the surface of every traffic route in a workplace shall be of a construction such that the floor or surface of the traffic route is suitable for the purpose for which it is used.
>
> (2) Without prejudice to the generality of paragraph (1), the requirements in that paragraph shall include requirements that—
> (a) the floor, or surface of the traffic route, shall have no hole or slope, or be uneven or slippery so as, in each case, to expose any person to a risk to his health or safety; and
> (b) every such floor shall have effective means of drainage where necessary.

[30] [2001] EWCA 1925.

(3) So far as is reasonably practicable, every floor in a workplace and the surface of every traffic route in a workplace shall be kept free from obstructions and from any article or substance which may cause a person to slip, trip or fall.
(4) In considering whether for the purposes of paragraph (2)(a) a hole or slope exposes any person to a risk to his health or safety—
(a) no account shall be taken of a hole where adequate measures have been taken to prevent a person falling; and
(b) account shall be taken of any handrail provided in connection with any slope.
(5) Suitable and sufficient handrails and, if appropriate, guards shall be provided on all traffic routes which are staircases except in circumstances in which a handrail cannot be provided without obstructing the traffic route.

The first point to be considered arises from sub-s (1) and the words 'shall be of a construction such that it is suitable for the purpose for which it is used'.

In *Palmer v Marks & Spencer plc*[31] the claimant tripped over a weather strip in a doorway. The issues in the case revolved around reg 12(1). The height of the weather strip was proud of the floor to the extent of some 8 to 9.5 mm. The judge at first instance found that the floor was of a suitable construction and that the weather strip presented no real danger. The Court of Appeal agreed and laid down general guidance as to how the court should approach the question of the words 'suitable for the purpose for which it is used'. The court said that whether a floor was of suitable construction was a matter of fact to be considered against the background of the use to which it was to be put and the other surrounding circumstances; the overall impression is that there are no absolute rules or duties and that the matter is one of impression for the judge hearing the case. The result in this case owes much to the fact that the weather strip had never caused anyone else a problem and perhaps more importantly the configuration is an extremely common one.

It may be that you will be faced with an argument that differential in height of only 8 to 9.5 mm does not make a floor unsafe. That is not what the Court of Appeal said in the *Palmer* case and it was very clear that whether it was unsafe had to be viewed from the point of view of whether or not there was any risk at all and that the extent of injury that might occur from a risk. It would not, it is contended, be open to an employer to have a situation, in a factory, where there was unevenness on the floor that was unnecessary and that served no useful purpose. Schiemann LJ put it aptly:

'In law context is everything. The context here is a shop, with it being expected that many people with varying degrees of physical mobility, in varying foot wear, and varying degrees of tiredness and attention, with varying amounts of bags and so on their person will use this floor to walk

---

[31] [2001] All ER (D) 123 (Oct).

> on ...The sort of slight rise we have here occurs everywhere ...the ordinary person would not regard ... such a floor ...as exposing him to a risk to his health and safety. Nor would the employer when constructing the floor.'

Paragraph 91 of ACOP provides as follows

> 'Holes, bumps or uneven areas resulting from damage or wear and tear which may cause a person to trip or fall should be made good.'......'Surfaces with small holes, for example metal gratings, are acceptable provided they are not likely to be a hazard.'

It is noteworthy that the ACOP does not call for perfection – and the action point arises when the irregularity may cause an accident to occur. Further, and importantly, the word may set the risk at a relatively low level.

When considering suitability of the construction of a floor there are a number of things to look at, such as the need for a step on a traffic routs. In *Lowles v The Home Office*[32] Mrs Lowles was injured after walking up a ramp and then tripping and falling over an unmarked step on her way into work. The Court of Appeal confirmed a breach of the absolute duty under reg 12(1). Dismissing the defendants' appeal Mance LJ (as he then was) said in para 14:

> 'In my view it could also have been relevant under regulation 12(1) for the Recorder to consider whether there was any good reason for having or maintaining the threshold ...'

He continued in paragraph 15:

> 'I do not see any reason why the unexpected threshold, constituting a step of unusual intermediate height at the top of a ramp, for which there was no apparent reason, should not in law be regarded as an obstruction in the floor or the surface of the traffic route.'

Regulation 12 was considered more recently, and in considerable depth, by the Court of Appeal in *Ellis v Bristol County Council*.[33] The facts in *Ellis* were that the accident occurred in a Nursing Home. A number of patients were known to be incontinent and to urinate in the corridors. The employers were well aware of this problem and had taken steps to lay mats, warn employees of the dangers and to be on their look out for urine on the floor. The claimant had just passed the end of a strategically placed mat when she slipped and fell.

The approach in *Palmer* was confirmed but the Court had to extend the analysis because *Palmer* only dealt with permanent features of floors in the workplace.

---

[32] [2004] EWCA CIV 985.

[33] [2007] EWCA Civ 685.

The Court of Appeal said that consideration had to be given to the suitability of the floor in the context of the use of which it was put and those provisions in reg 12(1) and (2) had to be considered together to decide whether or not there had been a breach.

In the Court of Appeal it was argued that the floor was properly constructed and therefore there could not be a breach of reg 12(1) or (2) and that therefore the claimant should fail. The Court of Appeal rejected that and came to the view, looking at the matter as a whole, that the floor was not of a sufficient or suitable construction for the purpose of which it was used at that time. It is important to note that the court said that it was the change brought about by the frequency of the urination that rendered the floor unsuitable and therefore unsafe; before that started to happen the floor was suitable.

5.7.1 Shall be kept free from obstructions (reg 12(3))

It is often forgotten in such cases that prevention is better than cure. Insurers will raise the point that the employer had a system of cleaning up spillages as and when they were found. This is a fallacy belied by the words 'shall be kept free'.

Once you see that sentiment in a letter you should immediately start considering how the spillage or obstruction got there in the first place. In nearly every case something will be left because there has been a system or human failure. The employer is vicariously liable for a breach of statutory duty caused by another employee. In cases, for example, involving accidents in kitchens, schools and other places, a not uncommon situation is that the floor is contaminated by food being dropped onto it. It is then necessary to consider how it might have got there and how that could have been avoided. For example, a potato chip on the floor may have fallen off the counter or from a bowl which was overloaded. If you can establish how things were done, then there is scope for discovering the most likely reason for the chip being on the floor.

This notion is illustrated in the case of *Burgess v Plymouth City Council.*[34] The claimant was a cleaner and went into a class room and tripped over a box used for storing lunch boxes. The defendants contested the matter mainly on the basis that the claimant ought to have been looking where she was going and it was her job to clean the classroom and therefore there could not be a breach of the statutory duties. The answer in this case, and the Court of Appeal said it was of general application, was that there was a system for storing such boxes, the box had not be stored, and therefore it had no business being on the floor; there was a breach of statutory duty albeit that the level of contributory negligence was high at 50%.

---

[34] [2005] EWCA Civ 1659.

There will be cases were spillages are almost inevitable and an accepted part of the process. Indeed, in some industrial situations it may be necessary to use the floor as the only place where residues can go. In such places the slip resistance of the floor must be increased to comply with the reg 12(1) duty. A simple comparative image can be conjured by considering the raised flooring seen around a swimming pool for example, which can be a workplace, and the shiny surfaces on route to the boardroom in an office building. Often the only measure taken is putting up a warning sign. Is that enough to deal with the reversed burden of proof? In *Cochrane v Gaughan*[35] Lord Brodie accepted that a warning sign was appropriately positioned and that there was no evidence of slipping on any occasion other than at the time of Carol Cochrane's accident. Even so the judge did not accept that such evidence was sufficient to make out the defence of reasonable practicability, in the absence of any evidence that it was all that could practicably be done.

5.8 CONTRIBUTION IN SLIPPING CASES

In *Palmer* Schiemann LJ identified a number of factors that come into play in assessing contribution ie tiredness, footwear, bags or other items being carried. The factual situation in *Ellis* is an excellent case study in the way in which judges should approach the assessment. In *Ellis* the Court of Appeal assessed contribution at a third. In particular the fact that the claimant had been warned that the floor might be slippery weighed heavily against her. In addition she was going about her usual duties and was not carrying anything. The outcome might have been different if she had been rushing to an emergency or an urgent response to a distressed resident at the home; in such circumstances the imperative of the warning by the employer would have been displaced and the actions might have been adjudged to have been inadvertence or at least scored at a lower rate of culpability on the claimant's part.

Therefore, in many cases involving reg 12 the issue of primary liability may not be difficult to overcome and there is a real need to concentrate on the issue of contribution.

5.9 TRAFFIC ROUTES

This is another important regulation in terms of practical day-to-day work on accidents at work and, following the discussion of the main duties, is best dealt with by a short disquisition after setting out the provision.

---

35 [2004] ScotCS 62.

Organisation etc. of traffic routes

17.—(1) Every workplace shall be organised in such a way that pedestrians and vehicles can circulate in a safe manner.

(2) Traffic routes in a workplace shall be suitable for the persons or vehicles using them, sufficient in number, in suitable positions and of sufficient size.

(3) Without prejudice to the generality of paragraph (2), traffic routes shall not satisfy the requirements of that paragraph unless suitable measures are taken to ensure that—
(a) pedestrians or, as the case may be, vehicles may use a traffic route without causing danger to the health or safety of persons at work near it;
(b) there is sufficient separation of any traffic route for vehicles from doors or gates or from traffic routes for pedestrians which lead onto it; and
(c) where vehicles and pedestrians use the same traffic route, there is sufficient separation between them.

(4) All traffic routes shall be suitably indicated where necessary for reasons of health or safety.

At first glance the regulation might be considered to be concerned with keeping vehicles and pedestrians separated. However, the use of the word or in sub-para (2) makes it clear that the safety aspects of the two must be considered separately.

The duty to ensure safe circulation of workers was considered in *Nichols v Beck Electronics*.[36] There the claimant was unaware that a door had been tied shut, for safety reasons, and was injured when he tried to open it; and there was found to be a breach of reg 17(2) as the route was not suitable (safe). In *Wallis v Balfour Beatty Rail Maintenance*[37] a locked gate barring access was found to be a breach; but on the facts was not causative of the injury.

The word 'suitable' is considered above in the discussion of the *Palmer* case. In the unreported case of *Pratt v Intermet Refractories*[38] it was said that a slight obstruction into a traffic route did not constitute a breach of duty. However, it is difficult to see how an obstruction can do other than make the route unsafe and therefore unsuitable. There is a danger in assessing the suitability of a traffic route, of falling into the trap of looking at what is reasonably safe rather than what is safe.

[36] (unreported) 30 June 2004, Norwich CC.
[37] [2003] EWCA Civ 72.
[38] (unreported) 21 January 2000.

5.10 WORKSTATIONS

Regulation 11(1) requires workstations to be arranged so that they are suitable for any person at work. The word suitable means safe and furthermore the regulation goes on to state that it must be suitable (safe) for *any* person likely to work at it in respect of any work that is likely to be done there. Again this will be a question of fact and degree but the important point to note when investigating is that the regulation personalises the workstation to the worker.

The section and ACOP are likely to be relevant to repetitive strain injury cases and need to be read in conjunction with the Health and Safety (Display Screen Equipment) Regulations 1992.[39]

5.10.1 Outdoor Workstations

Firstly, an outdoor workstation has to be arranged that, in so far as it is reasonably practicable, it provides protection from adverse weather.[40]

Secondly, and without any such limitation, the workstation must allow any person to leave it swiftly or, as appropriate to be assisted to leave, in the event of emergency. Further, the workstation must be arranged so that it ensures that any person at the workstation is not likely to slip or fall and this produces a higher duty than that under reg 12(3).

5.11 SEATING

Regulation 11(4) requires that seating provided for any person must be suitable for that person. In addition if a suitable foot rest is needed that must also be suitable for that person. Again, suitable and necessary are probably to be understood in terms of health and safety. Since 17 September 2002 the duty is extended to seating provided in restrooms. The regulation is more likely to be of relevance in a progressive or repetitive disease case. If a chair breaks the claim can be brought under the absolute duty in the Equipment Regulations.

5.12 LIGHTING

This is governed by reg 8 which requires that every workplace shall have suitable and sufficient lighting. The Factories Act 1961 contained a similar provision in s 5: 'effective provision shall be made for securing and maintaining suitable and sufficient lighting.' In *Thornton v Fisher &*

---

[39] SI 1992/2792.

[40] SI 1992/2792, reg 11(3).

Ludlow Ltd[41] and *Davies v Massey Ferguson Perkins Ltd*[42] it was held that the duty is strict – thus if a light is provided and not turned on or suddenly fails then there is a breach.

However, whether the lighting provided is suitable and sufficient is a matter of fact and degree. In *Miller v Perth & Kinross Council* it was held that an area in shadow on a sloping pavement as a result of nearby street lighting constituted a breach of reg 8. There is an element of fact and degree to be considered and in such a case evidence of complaints or previous accidents maybe important. An earlier Court of Appeal case, *Lane v Gloucestershire Engineering*[43] was not mentioned in *Miller* – there the judge's finding that an area in shadow did not create a breach where the overall lighting was sufficient. The approach to the problem in *Lane* was arguably wrong and the case is an illustration of the Appeal Court's reluctance to interfere with findings of fact. However, it illustrates that when the word 'suitable' or 'sufficient' appears in the legislation there is a danger of assessing what is reasonably safe rather than whether there is an unsafe situation.

5.13 VENTILATION

Regulation 6 provides that effective and suitable provision should be made to ensure that every enclosed workplace is ventilated by a sufficient quantity of fresh purified air. Where mechanical ventilation systems are used to this end they must be maintained in an efficient state and efficient working order and in good repair and if appropriate be subject to the suitable system of maintenance. If there is a failure of the ventilation plant this must be signalled by a visible audible warning.

This is a generalised regulation and adds nothing to the specific legislation that deals with hazardous substances.[44]

5.14 TEMPERATURE ETC FOR INSIDE WORKPLACES

Regulation 7 states that during working hours the temperature in all workplaces inside buildings shall be reasonable. In addition a method of heating or cooling shall not be used that results in the escape into a workplace of fumes, gas or vapour of such character and to such extent

---

41 [1968] 2 All ER 241.
42 [1986] ICR 580.
43 [1967] 2 All ER 293.
44 Control of Substances Hazardous to Health Regulations 1992, SI 2002/2677 (COSSH), which are outside the ambit of this work.

that they are likely to be injurious or offensive to any person.[45] From 17 September 1992 there is a requirement for adequate thermal insulation.[46]

There is no definition of reasonable and no maximum or minimum temperatures. The ACOP gives limited guidance suggesting a minimum of 16 degrees celsius or 13 degrees when the work involves severe physical effort.

Where there is excessive heat or severe cold then reliance will need to be placed on the Personal Protective Equipment Regulations.

5.15 CLEANLINESS

The duty, under reg 9, is to keep the workplace, furniture, furnishings and fittings sufficiently clean and ensure the surfaces of walls, floors and ceilings of all workplaces inside buildings are capable of being kept sufficiently clean. So far as it is reasonably practicable, waste materials should not be allowed to accumulate in any workplace, except in suitable receptacles.

There is an overlap here with regs 5, 12, 13, 20(2) (b) and 21(2) (g). However, it should be noted that the duties are very general and the use of the word 'sufficiently', twice, in the regulation, leaves a wide margin of appreciation or room for argument as to what is or is not sufficiently clean.

The ACOP deals mainly with the need to clean the workplace – but the standard will depend, it is said, on the use to which it is put.

The provision was considered in *Pratt v Intermet Refractories Ltd.*[47]

5.16 ROOM DIMENSIONS AND SPACE

This is governed by reg 10, which requires every room where a person works shall have sufficient floor area. In *Butler v Grampian University Hospitals NHS Trust*,[48] Butler was an outpatient assistant at her employer's hospital. She alleged that she suffered leg and back pain after bearing the weight of a disabled patient who fell while being assisted by Butler and a carer in a toilet cubicle. Butler alleged that the cubicle was unsuitable for wheelchair bound patients. Part of her case was that the floor area was inadequate in terms of reg 10(1). Lord Macfadyen

45 SI 1992/3004, reg 7(2).
46 Health and Safety (Miscellaneous Amendments) Regulations 2002, SI 2002/2174, reg 6(f) inserted reg 7(1A).
47 (unreported) 21 January 2000, CA.
48 2002 SLT 985.

accepted this approach and held that the sufficiency of a floor area under reg 10(1) could not be judged without reference to the work being undertaken.

Regulation 10(2) refers to Sch 1 Part 1, which is set out below:

> **'Space**
>
> **1.** No room in the workplace shall be so overcrowded as to cause risk to the health or safety of persons at work in it.
>
> **2.** Without prejudice to the generality of paragraph 1, the number of persons employed at a time in any workroom shall not be such that the amount of cubic space allowed for each is less than 11 cubic metres.
>
> **3.** In calculating for the purposes of this Part of this Schedule the amount of cubic space in any room no space more than 4.2 metres from the floor shall be taken into account and, where a room contains a gallery, the gallery shall be treated for the purposes of this Schedule as if it were partitioned off from the remainder of the room and formed a separate room.'

5.16.1 Confined Spaces

From 28 January 1998 working in confined spaces is governed by the Confined Spaces Regulations 1997.[49] The provisions deal with situations which are clearly potentially hazardous.

A confined space includes any chamber tank, vat, silo, pit, trench pipe, sewer flue, well or other similar spaces ,which by virtue of its enclosed nature gives rise to a reasonably foreseeable specified risk.[50]

A specified risk is one of:[51]

- serious injury to any person at work arising from fire or explosion;
- loss of consciousness of any person or work arising from an increase in body temperature
- loss of consciousness or asphyxiation of any person in work arising from gas fume vapour or the lack of oxygen;
- drowning of any person at work arising from any increase in the level of a liquid; or

---

49 SI 1997/1713 repealing Factories Act 1961, s 30.
50 SI 1997/1713, reg 1 (2).
51 SI 1997/1713, reg 1 (2).

- the asphyxiation of any person arising from a free-flowing solid or the inability to escape to a respirable due to entrapment by a free-flowing solid.

In common with other legislation passed at that time there is a hierarchy of avoidance measures.

Firstly, no one is to enter into a confined space to carry out work for any purpose unless it is not reasonably practicable to achieve that purpose without entering the confined space.

Secondly, so far as it is reasonably practicable, no person at work shall enter or carry out any work in or (other than as a result of an emergency) leave a confined space otherwise in accordance with a system of work which, in relation to any relevant specified risks to health, renders the work safe

Thirdly, before any work starts in a confined space suitable and sufficient arrangements must be in place to rescue a person in the event of an emergency.

5.17 WINDOWS AND SKYLIGHTS ETC

Regulations 14 and 15 are straightforward and are set out below. The language imposes a strict duty on the employer.

> **'Windows, and transparent or translucent doors, gates and walls**
>
> **14.**—(1) every window or other transparent or translucent surface in a wall or partition and every transparent or translucent surface in a door or gate shall, where necessary for reasons of health or safety—
> (a) be of safety material or be protected against breakage of the transparent or translucent material; and
> (b) be appropriately marked or incorporate features so as, in either case, to make it apparent.
>
> **Windows, skylights and ventilators**
>
> **15.**—(1) No window, skylight or ventilator which is capable of being opened shall be likely to be opened, closed or adjusted in a manner which exposes any person performing such operation to a risk to his health or safety.
>
> (2) No window, skylight or ventilator shall be in a position when open which is likely to expose any person in the workplace to a risk to his health or safety.
>
> **Ability to clean windows etc. safely**
>
> **16.**—(1) All windows and skylights in a workplace shall be of a design or be so constructed that they may be cleaned safely.

(2) In considering whether a window or skylight is of a design or so constructed as to comply with paragraph (1), account may be taken of equipment used in conjunction with the window or skylight or of devices fitted to the building.'

5.18 DOORS AND GATES

'Doors and gates

18.—(1) Doors and gates shall be suitably constructed (including being fitted with any necessary safety devices).

(2) Without prejudice to the generality of paragraph (1), doors and gates shall not comply with that paragraph unless—

(a) any sliding door or gate has a device to prevent it coming off its track during use;

(b) any upward opening door or gate has a device to prevent it falling back;

(c) any powered door or gate has suitable and effective features to prevent it causing injury by trapping any person;

(d) where necessary for reasons of health or safety, any powered door or gate can be operated manually unless it opens automatically if the power fails; and

(e) any door or gate which is capable of opening by being pushed from either side is of such a construction as to provide, when closed, a clear view of the space close to both sides.'

Regulation 18(1) lies down the general broad duty and the remainder deals with particular circumstances and types of door. It should also be noted that any transparent or translucent part of the door or gate must comply with reg 14: see above.

While a door must be suitably constructed imposes a strict duty – there is still room for argument as to whether the door is suitable.

Hurd v Stirling Group plc[52] is an illustration of the difficulty in a common type of case where the allegation is that the door closed too quickly, ie in this case one second.

The claimant was walking down a corridor (with some 75 other employees) when a plastic door closed and trapped her foot under it. The judge found as a fact that the gap at the foot of the door was 1 inch (the claimant alleged it was 3 inches) and that it had taken two seconds to close and the claimant's foot was caught because the door was flexible and had ridden up over her shoe. The trial judge, noting that the door had been used without incident for some years, rejected the allegation of breach of reg 18, holding that the door was suitably constructed. It was also submitted that the combination of the weight and speed of closure

---

[52] [1999] EWCA Civ 1487.

and the flexibility of the door gave rise to a foreseeable risk of injury – foreseeability is a matter of fact and the court rejected the argument. The Court of Appeal fully agreed with the judge. It is arguable that the fact that the claimant's foot was caught at all made the door unsuitable and that inattention on her part was excusable.

However, if the door is not suitable then liability is strict. In *Beck v United Closures and Plastics*[53] here the doors were heavy and formed part of the machinery. While they were therefore work equipment Lord McEwan also held that there was a breach of reg 18: the positioning of the handles made it a difficult task and therefore they were not suitable.

5.19 ESCALATORS AND MOVING WALKWAYS

'Escalators and moving walkways

19. Escalators and moving walkways shall:—

(a) function safely;
(b) be equipped with any necessary safety devices;
(c) be fitted with one or more emergency stop controls which are easily identifiable and readily accessible.'

The duties are clearly strict and a malfunction will impose liability. The words 'function safely' impose a continuing obligation and whether or not there are safety devices will be irrelevant in establishing liability because any injury is likely to arise as a result of the failure to function safely.

5.20 WELFARE PROVISIONS

Regulations 20 to 24 contain detailed provisions relating to comfort, hygiene and rest facilities at work. They will rarely be relied upon in accident claims but will be relevant in cases of industrial disease, especially reg 21 that should be considered in dermatitis cases.

'Sanitary conveniences

20.—(1) Suitable and sufficient sanitary conveniences shall be provided at readily accessible places.

(2) Without prejudice to the generality of paragraph (1), sanitary conveniences shall not be suitable unless—

(a) the rooms containing them are adequately ventilated and lit;
(b) they and the rooms containing them are kept in a clean and orderly condition; and
(c) separate rooms containing conveniences are provided for men and women except where and so far as each convenience is in a separate room the door of which is capable of being secured from inside.

[53] 2001 SLT 1299.

(3) It shall be sufficient compliance with the requirement in paragraph (1) to provide sufficient sanitary conveniences in a workplace which is not a new workplace, a modification, an extension or a conversion and which, immediately before this regulation came into force in respect of it, was subject to the provisions of the Factories Act 1961, if sanitary conveniences are provided in accordance with the provisions of Part II of Schedule 1.

Washing facilities

21.—(1) Suitable and sufficient washing facilities, including showers if required by the nature of the work or for health reasons, shall be provided at readily accessible places.

(2) Without prejudice to the generality of paragraph (1), washing facilities shall not be suitable unless—
(a) they are provided in the immediate vicinity of every sanitary convenience, whether or not provided elsewhere as well;
(b) they are provided in the vicinity of any changing rooms required by these Regulations, whether or not provided elsewhere as well;
(c) they include a supply of clean hot and cold, or warm, water (which shall be running water so far as is practicable);
(d) they include soap or other suitable means of cleaning;
(e) they include towels or other suitable means of drying;
(f) the rooms containing them are sufficiently ventilated and lit;
(g) they and the rooms containing them are kept in a clean and orderly condition; and
(h) separate facilities are provided for men and women, except where and so far as they are provided in a room the door of which is capable of being secured from inside and the facilities in each such room are intended to be used by only one person at a time.
(3) Paragraph (2)(h) shall not apply to facilities which are provided for washing hands, forearms and face only.

Drinking water

22.—(1) An adequate supply of wholesome drinking water shall be provided for all persons at work in the workplace.

(2) Every supply of drinking water required by paragraph (1) shall—
(a) be readily accessible at suitable places; and
(b) be conspicuously marked by an appropriate sign where necessary for reasons of health or safety.
(3) Where a supply of drinking water is required by paragraph (1), there shall also be provided a sufficient number of suitable cups or other drinking vessels unless the supply of drinking water is in a jet from which persons can drink easily.

Accommodation for clothing

23.—(1) Suitable and sufficient accommodation shall be provided—
(a) for the clothing of any person at work which is not worn during working hours; and

(b) for special clothing which is worn by any person at work but which is not taken home.

(2) Without prejudice to the generality of paragraph (1), the accommodation mentioned in that paragraph shall not be suitable unless—

(a) where facilities to change clothing are required by regulation 24, it provides suitable security for the clothing mentioned in paragraph (1)(a);

(b) where necessary to avoid risks to health or damage to the clothing, it includes separate accommodation for clothing worn at work and for other clothing;

(c) so far as is reasonably practicable, it allows or includes facilities for drying clothing; and

(d) it is in a suitable location.

Facilities for changing clothing

24.—(1) Suitable and sufficient facilities shall be provided for any person at work in the workplace to change clothing in all cases where—

(a) the person has to wear special clothing for the purpose of work; and

(b) the person cannot, for reasons of health or propriety, be expected to change in another room.

(2) Without prejudice to the generality of paragraph (1), the facilities mentioned in that paragraph shall not be suitable unless they include separate facilities for, or separate use of facilities by, men and women where necessary for reasons of propriety.

Facilities for rest and to eat meals

25.—(1) Suitable and sufficient rest facilities shall be provided at readily accessible places.

(2) Rest facilities provided by virtue of paragraph (1) shall—

(a) where necessary for reasons of health or safety include, in the case of a new workplace, an extension or a conversion, rest facilities provided in one or more rest rooms, or, in other cases, in rest rooms or rest areas;

(b) include suitable facilities to eat meals where food eaten in the workplace would otherwise be likely to become contaminated.

(3) Rest rooms and rest areas shall include suitable arrangements to protect non-smokers from discomfort caused by tobacco smoke.

(4) Suitable facilities shall be provided for any person at work who is a pregnant woman or nursing mother to rest.

(5) Suitable and sufficient facilities shall be provided for persons at work to eat meals where meals are regularly eaten in the workplace.

25A. Where necessary, those parts of the workplace (including in particular doors, passageways, stairs, showers, washbasins, lavatories and workstations) used or occupied directly by disabled persons at work shall be organised to take account of such persons.'[54]

---

[54] Amended by Health Safety (Miscellaneous Amendments) Regulations 2002, SI 2002/2174.

CHAPTER 6

WORKING AT HEIGHT

6.1 INTRODUCTION

The Work at Height Regulations 2005[1]came in to force on 6 April 2005 and gave effect to Council Directive 2001/45/EC. One of the aims of the Regulations was to bring together all requirements for safe working at height in existing legislation in a single set of regulations applicable to all industries. The old statutory regime is of largely historical interest and therefore this chapter will concentrate on the new rather than the old regime.

The chapter does not follow the Regulations in sequential order. The first part deals with the general principles applicable to safe work practices to be followed and applied at work. Then special consideration is given to work with ladders – one of the most common type of cases that the practitioner will encounter.

It is axiomatic that working at height is dangerous and indeed the activity accounts for more fatalities and serious injuries than any other carried out at work. Putting the matter into context it should be noted that in 2003/2004, 67 people died and nearly 4,000 suffered serious injury as a result of a fall from height in the workplace and that this accounted for some 15% of all major injures. It is not surprising therefore that working at height is a high priority programme area for the HSE. The latest HSE figures show that 45 people died from a fall from height at work in 2006/07. This was a record low. But the HSE confirm that falls from height remain the most common kind of accident causing fatal injuries.

This is not surprising as often on the street you will have observed the precarious approach to working at height with long extension ladders to clean windows several storeys above the ground. Now you will, hopefully, see extended poles with brushes at the end. The case of *General Cleaning Contractors v Christmas*[2] describes how windows used to be cleaned by standing on thin ledges – and the tenor of the judgments appears to indicate that this was almost acceptable! The advent of cherry pickers and

---

1 SI 2005/735.

2 [1953] AC 180.

scissor lifts make such methods obsolete; the world has to a large extent moved on but there will still be instances of dangerous practices borne of the desire to cut corners or to save a few pounds.

6.2 MAJOR CHANGES

There are three major points that should be to the forefront of the thinking for practitioners especially those who have been accustomed to the old rules.

6.2.1 Risk Assessments

The Regulations are underpinned by the need for organisation and planning (risk assessment) and the use of competent employees. The centre piece of the 2005 Regulations is reg 6[3] which imposes an absolute duty to ensure that all work at height is subject to a risk assessment. Previously, reliance had to be placed upon the general duty to carry out risk assessments under reg 3 of the Management Regulations; the new provision relating to work at height should finally dispel any argument from insurers that risk assessments are an optional extra. In addition practitioners should immediately look for the assessment or lack of it. The failure to risk assess will create a breach of duty and will be *the* documentation to look for at an early stage of investigation.

6.2.2 The Two-Metre Rule

This change from previous legislation is worthy of a particular note. It was previously the case that measures in relation to working at height in construction cases[4] were governed by the two-metre rule. However, in carrying out research the HSE discovered that 60% of major injuries were caused by falls from heights of less than two metres. It was therefore decided to abandon that as a measure at which special precautions had to be taken. (The last remnant of the two-metre rule is found in reg 12, which requires in sub-para (4) that where a working platform is used for construction work and it is the place from which a person can fall two metres or more it must be inspected at least every seven days.)

6.2.3 Inspections

Regulation 12 of the 2005 Regulations[5] sets up a system for inspection. When investigating a claim the documentation generated, or that should have been generated, will be of importance.

---

3 See **6.5.3** below.

4 Construction (Health, Safety and Welfare) Regulations 1996, SI 1996/1592, reg 6. Precautions were to be taken where a fall of two metres or more was likely to cause injury.

5 See **6.5.9** below

6.3 APPLICATION

6.3.1 Control

Regulation 3(2) of the 2005 Regulations states that the requirements imposed by these regulations on an employer shall apply in relation to work (a) carried out by an employee of his as well as (b) any other person under his control to the extent of his control. Subsequently each of the regulations is framed on the basis that the employer shall conform to the duty. The effect of the wording is therefore that the employer cannot delegate the duty to another person and the employer will always have a liability under the Regulations, to his own employees, even if he is not on site and has no control over the safety aspects.

The duties are also imposed, as with other modern safety regulations, on those who have control. The concept of control, in this context, is discussed elsewhere.

6.3.2 Exceptions to Application

Regulation 3 of the 2005 Regulations removes form the ambit of the regulations masters' crews and employers in respect of normal ship board activities to the extent that the activities are carried out solely by the crew under the direction of the master, provided that there is no likelihood that there will be a danger to the master or crew.

There are exemptions in respect of normal activities on board a ship, dock works, loading and unloading of fishing vessels and offshore installations.[6]

One problem that arose in respect of application, before the introduction of the Regulations, was in respect of caving and climbing. Therefore, the provision of instruction or leadership to one of more persons in connection with their engagement in caving or climbing by way of sport, recreation, team building or similar activities[7] was excluded. However, this was dealt with in the Work at Height (Amendment) Regulations 2007[8] which inserts a new reg 14A into the 2005 Regulations. The wording of reg 14A is tortuous and is therefore set out in full.

> **'14A. Special provision in relation to caving and climbing**
>
> (1) Paragraph (2) applies in relation to the application of these Regulations to work concerning the provision of instruction or

---

6 See SI 2005/735, reg 3(4)(a)-(c).

7 SI 2005/735, reg 4 (3)(d).

8 SI 2007/117.

leadership to one or more persons in connection with their engagement in caving or climbing by way of sport, recreation, team building or similar activities.

(2) Where this paragraph applies, an employer, self-employed person or other person shall be taken to have complied with the caving and climbing requirements, if, by alternative means to any requirement of those requirements, he maintains in relation to a person at such work as is referred to in paragraph (1) a level of safety equivalent to that required by those requirements.

(3) For the purposes of paragraph (2), in determining whether an equivalent level of safety is maintained, regard shall be had to —

(a) the nature of the activity;

(b) any publicly available and generally accepted procedures for the activity; and

(c) any other relevant circumstances.

(4) In this regulation —

(a) "caving" includes the exploration of parts of mines which are no longer worked;

(b) "climbing" means climbing, traversing, abseiling or scrambling over natural terrain or man-made structures; and

(c) "the caving and climbing requirements" means regulation 8(d)(ii), so far as it relates to paragraph 1 in Part 3 of Schedule 5, and that paragraph.".

(5) In paragraph 1 of Part 3 of Schedule 5 —

(a) at the beginning, insert "Except as provided in paragraph 3,"; and

(b) in sub paragraph (a), omit "subject to paragraph 3'.'

The meaning appears to be that the safety of any particular caving or climbing activity is to be judged by publicly available and generally accepted procedures for the activity or by ensuring that the (equivalent) standard is met by alternative means.

One problem is that the employer's duties under the Regulations are non delegable and the wording of reg 14A appears to maintain that position if an employee is sent on a team-building exercise. Is the duty complied with by selecting a reputable provider for the activity? It is arguable that if the provider departs from generally accepted practice the non-delegable nature of the duty cannot be avoided. Alternatively, the words 'by alternative means to any requirement of those requirements, he maintains in relation to a person at such work as is referred to in paragraph (1) a level of safety equivalent to that required by those requirements' may be an escape route for the employer by allowing for selection of a competent provider.

6.4 WORK AT HEIGHT

Is defined in reg 2 of the 2005 Regulations as

(a) work at any place, including at or below ground level;

(b) obtaining access to or egress from such place while at work, except by a staircase in a permanent workplace; where, if measures required by these Regulations were not taken, a person could fall a distance liable to cause personal injury.

The term 'fall' was considered in the case of *Campbell v East Renfrewshire Council*.[9] The claimant was working on a steep embankment at a roundabout in the wind and rain, slipped, and rolled about 20 feet to the bottom of the embankment. He claimed that the embankment was unsafe in terms of reg 13(1)(a) of the Workplace (Health, Safety and Welfare) Regulations 1992 and that the employer ought to have taken precautions to prevent any person from 'falling a distance likely to cause personal injury'.

The employers countered that 'falling a distance' in reg 13 of the 1992 regulations was intended to cover falling from a height, rather than falling over at ground level. Although Campbell rolled down a slope, he remained on the same surface, and did not 'fall a distance'. Campbell argued that 'falling' was synonymous with 'sliding', 'tumbling' or 'rolling', and that he had ended up at a lower level than that from which he had fallen. The judge held that the claimant had not 'fallen a distance', within meaning of reg 13 of the 1992 regulations, since he only fell onto the ground on which he was standing. Regulation 13 was intended to prevent people falling from one surface onto a lower surface, and sustaining injury as a result of the distance between those surfaces.

It is questionable whether the decision is correct. If the claimant had been working on a sloping roof and slipped down to the eaves and suffered injury, without going over the edge, then it is difficult to see an arguable defence.

6.5 ORGANISATION AND PLANNING

Regulations 4, 5, 6 and 7 of the 2005 Regulations set the scene for the modern approach to safety when working at height. One of the hallmarks of the six pack legislation, introduced in 1992, was the emphasis put upon the management of risk and proactive planning by the employer. Regulation 6 is perhaps the high watermark, thus far, of the way in which an employer should now approach the task of ensuring safety at work.

6.5.1 Regulation 4

Under reg 4 of the 2005 Regulations, every employer must ensure that the work is properly planned, appropriately supervised and carried out in a manner which is so far as is reasonably practicable safe and that the

[9] (unreported) 31 March 2004, OH Court of Session (Temporary Judge R F MacDonald) A3090/01.

planning includes the selection of work equipment in accordance with reg 7. The planning must also take into account emergencies and rescue.

Regulation 4(3) requires that work at height must only be carried out when the weather conditions do not jeopardise the safety of persons involved in the work. However, this does not apply when the emergency services are involved; in this respect the police, fire and ambulance services are mentioned specifically.

6.5.2 Regulation 5

Under reg 5 of the 2005 Regulations, the employer must ensure that those working at height or planning the work are competent. In respect of trainees the supervisor must be competent.

This is an important provision because often in such cases the employer will defend on the basis that the claimant is experienced in the work. However, experience and competency are not the same thing. In such cases there will be a need to consider the way in which work has been done in the past. In many companies especially smaller ones training is very much on the job and if the worker has been led into bad practices by example then it is arguable that the worker is not competent. If that is the case then there will be a breach of duty by the employer.

There are various industry training schemes and consideration should be given to investigating if and when the employee has received such training. In addition such training should be reinforced by refreshers and evidence relating to tool box talks may also be useful.

6.5.3 Regulation 6

Regulation 6(1) of the 2005 Regulations specifically states that the employer shall take account of a risk assessment under reg 3 of the Management Regulations.

The first step to be taken by the employer is to question whether or not it is reasonably practicable to carry out the work safely otherwise that at height.[10]

If the work is to be carried out at height then suitable and sufficient measures to prevent, so far as it is reasonably practicable, any person falling a distance liable to cause personal injury must be taken.[11]

Regulation 6(4) then goes on to define the steps that the employer must take to comply with reg 6(3), as follows:

---

[10] SI 2005/735, reg 6(2).

[11] SI 2005/735, reg 6(3).

(1) The question to be asked is whether or not the work can be carried out from the existing place of work or using an existing means of access or egress which complies with Sch 1.

(2) If it is not reasonably practicable then the employer must provide sufficient work equipment to prevent, so far as it is reasonably practicable, a fall occurring.

(3) If the work is not to be carried out from the existing place of work, or means of access or egress, then further steps must be taken if there is still a risk of falling by way of providing, so far as it is reasonably practicable, sufficient work equipment to minimise the distance and consequences of a fall.

(4) Where it is not reasonably practicable to minimise the distance, the consequences of a fall must be minimised.

The longstop in the equation is that the employer must provide additional training and instruction or take other suitable and sufficient measures to prevent, so far as it is reasonable practicable, any person falling a distance liable to cause personal injury. This provision is unlikely to come into play because it envisages a situation where having taken all the above precautions there is still a risk of falling.

6.5.4 Regulation 7

Regulation 7 of the 2005 Regulations is concerned with the selection of work equipment to be used while working at height and covers not just tools but equipment to prevent falls.

The primary rule is that collective safety measures are to be preferred to individual or personal measures. This means that harnesses etc are regarded as lesser means to the end of safe working than a scaffold or other means of safeguarding the safety of the workers.

The regulation goes on to list the considerations that must be taken on board by the employer to ensure safety:

(1) the working conditions and the risks to the safety of persons at the place where the work equipment is to be used;

(2) in the case of work equipment for access and egress, the distance to be negotiated;

(3) the distance and consequences of a potential fall;

(4) the duration and frequency of use;

(5) the need for easy and timely evacuation and rescue in an emergency;

(6) any additional risk posed by the use, installation or removal of that work equipment or by evacuation and rescue from it; and

(7) the other provisions of these Regulations.

The list can be adopted in investigating and evaluating the merits of a claim to establish whether the employer has approached the question in a through manner.

6.5.5 Regulation 8 and the Schedules

Regulation 8 of the 2005 Regulations sets out the requirements for employers in relation to particular work equipment, with the specific details contained in Schs 2-6. The schedules are considered below at **7.6**.

6.5.6 Regulation 9

Many factories have glass sections to otherwise fragile surfaces in the roof that present a dangerous situation. A fragile surface is defined as a surface liable to fail if any reasonably foreseeable load is placed upon it.[12]

The design of reg 9 of the 2005 Regulations is as follows.

(1) Firstly, so far as is reasonably practicable where work can be done safely and ergonomically no person should for any reason get close to or work on the area of danger.[13]

(2) Secondly, if that cannot be avoided then the area must be covered or supported so that any foreseeable load placed on the covering is borne by the covering or support.[14]

(3) Thirdly, where any one may get near the surface prominent warning notices must be put in place or, where that is not reasonable practicable, such persons must be made aware of the danger by other means.[15]

The provisions in respect of signing, but not the other provisions, do not apply to attendance by the emergency services including the police fire and ambulance services.[16]

---

12 SI 2005/735, reg 2.
13 SI 2005/735, reg 9(1).
14 SI 2005/735, reg 9(2).
15 SI 2005/735, reg 9(3).
16 SI 2005/735, reg 9(4).

6.5.7 Regulation 10

Regulation 10 of the 2005 Regulations deals with falling objects. The primary duty is to take suitable and sufficient steps to prevent the fall of any object[17] and, where that is not reasonably practicable, then suitable and sufficient steps must be taken to prevent any person being struck by a falling object.[18]

There is a specific prohibition placed on the employer on the throwing or tipping of any object liable to a cause personal injury. This appears to be an absolute duty that the employer cannot escape from if an employee decides to act in such a way.[19]

A regular problem is in respect of wind-borne items and this is covered by imposing the duty to ensure that materials and objects are stored in such a way so as to prevent the risk of injury to any person from the collapse or unintended movement of such. Again, the duty is absolute.[20]

The Construction (Head Protection) Regulations 1989,[21] as amended by the Personal Protective Equipment Regulations 1992, apply to construction sites. Other work places will be governed by the Personal Protective Equipment Regulations the provisions of which are dealt with in Chapter 8.

The 1989 Regulations impose a non-delegable duty on the employer to ensure that head protection is made available. In addition the employer must ensure that the head protection is worn unless there is no foreseeable risk and the same duty is imposed upon those in control of the work.[22]

The regulations require the person in control to ensure that hard hat zones are designated. While the controller has the primary duty employers and those in control of workers also have joint responsibility.[23] The site controller is responsible for establishing rules concerning the use of hard hats on site.[24]

The regulations also impose a duty on employees and the self employed to comply with the rules in respect of the wearing of hats.[25]

---

17 SI 2005/735, reg 10(1).
18 SI 2005/735, reg 10(2).
19 SI 2005/735, reg 10(3).
20 SI 2005/735, reg 10(4).
21 SI 1989/2209.
22 SI 1989/2209, reg 4.
23 SI 1989/2209, reg 5.
24 SI 1989/2209, reg 6.
25 SI 1989/2209, reg 6.

It is worth noting that a hard hat will not necessarily protect the worker from injury if an object falls from a considerable height – in other words the failure to do so would not have or might not have prevented the injury.

6.5.8 Regulation 11

Under reg 11 of the 2005 Regulations, where there is a risk of a person or an object falling that is liable to cause a person entering the area injury, that area of the work place must be equipped with devices to prevent unauthorised access to it., If it is not reasonably practicable to have such devices, signs to clearly indicate the area of a danger.

6.5.9 Regulation 12

Regulation 12 of the 2005 Regulations imposes the duty to ensure that work equipment that is dependent for its safe use on how it is installed or assembled is inspected in the position in which it is to be used prior to commencement of its use.[26]

Where work equipment is exposed to conditions causing deterioration liable to result in a dangerous situation it has to be inspected at suitable intervals. Each time exceptional circumstances occur that are liable to jeopardise its safety steps must be taken to ensure that health and safety conditions are maintained and that any deterioration can be detected and remedied in good time. This provision is aimed at ensuring the integrity of the equipment, for example after a storm involving high winds.[27]

Furthermore, if a working platform is used for construction work and is in a position where a person could fall 2 metres or more then it is not to be used unless it has been inspected in the position in which it is being used within the previous seven days.[28]

Equipment that is not covered by the Lifting Operations and Lifting Equipment Regulations 1998[29] must be inspected before it leaves the employer's undertaking or if obtained from a third person, unless it is accompanied by physical evidence that the last required inspection to be carried out has been completed.[30] Further, a report, complying with Sch 7 must be prepared and provided within 24 hours of completion to the person on whose behalf the inspection has been carried out.[31] There are further provisions relating to the keeping and delivering of records of

---

[26] SI 2005/735, reg 12(2).
[27] SI 2005/735, reg 12(3).
[28] SI 2005/735, reg 12(4).
[29] SI 1998/2307.
[30] SI 2005/735, reg 10(6)-(8).
[31] SI 2005/735, reg 12(5).

inspection.[32] There are special provisions in relation to lifting operations and lifting equipment that are covered under those regulations.

6.5.10 Regulation 13

Regulation 13 of the 2005 Regulations requires that every workplace where work is carried out must be checked on each occasion before the place is used. This includes the surface and every permanent rail or other such form of protection measure that is used.

6.5.11 Regulation 14

Regulation 14 of the 2005 Regulations imposes a duty on any person at work to report to the person who controls his work any defect or activity relating to work at height that he knows is likely to endanger his or another person's safety.[33]

Additionally there is a requirement to use any work equipment or safety device provided to him for work at height by his employer, or by a person under whose control he works, in accordance with any training or relevant instruction given by his employer or controller.

This is likely to be used to argue contribution on the part of the worker. However, it is submitted that where the employer/controller knows of the problem already then there is no duty on the worker to make the report. Alternatively, the failure to do can arguably be said to have been causative of the injury.

6.5.12 Regulation 15

Regulation 15 of the 2005 Regulations provides the HSE with power to permit exemptions from the provisions relating to para 3 of Sch 2 which relate to the height of guard rails and the provision of toe boards.

6.5.13 Regulation 16

Under reg 16 of the 2005 Regulations the Secretary of State has power to exempt the armed forces from the regulations in the interests of national security.

---

32 SI 2005/735, reg 12(6)-(8).

33 SI 2005/735, reg 14(1).

6.6 THE SCHEDULES

6.6.1 Schedule 1 Requirements for existing places of work and means of access or egress of height.

Under Sch 1 to the 2005 Regulations, such places must be stable and of sufficient strength and rigidity for its intended purpose and rest upon a stable and sufficiently strong surface.

The safety of passage of persons must be ensured by providing that the area is one of sufficient dimensions. In this respect regard must be had to the safe use of any plant or materials that will be required to be used thereupon and the area must be safe having regard to the work to be carried out there. Falls must be prevented by suitable and sufficient means.

There is then an absolute duty to ensure that the place is constructed and used and maintained so as to prevent so far as it is reasonably practicable the risk of slipping or tripping of any person being caught between it and any adjacent structure. If there are moving parts appropriate devices must be installed to prevent inadvertent movement during work at height.

6.6.2 Schedule 2 Requirements for guard rails, toe boards, barriers and similar collective means of protection

Under Sch 2 to the 2005 Regulations, means of protection must be of sufficient strength and rigidity for the purpose for which they are used; this imposes an absolute duty.[34] So far as is reasonably practicable means of protection must be secured and used so that they cannot be accidentally displaced;[35] and to prevent the fall of any person or object from the place of work.[36]

The schedule contains specific rules in respect of construction work and where it is being carried out the top of any guard rail must be at least 950 mm above the edge from which any person is liable to fall. Intermediate guard rails must be placed so that the gap is not more than 470 mm between it and any other means of protection.[37]

It is not unusual to secure scaffold to another object and, in such circumstances, the object must be of sufficient strength and suitable for that purpose.[38]

---

[34] SI 2005/735, Sch 2, para 2(a).
[35] SI 2005/735, Sch 2, para 2(b).
[36] SI 2005/735, Sch 2, para 2(c).
[37] SI 2005/735, Sch 2, para 3(c).
[38] SI 2005/735, Sch 2, para 4.

The overall objective is therefore to provide a secure working environment designed to keep people and objects within its parameters. However, there will be occasions when an opening is necessary for means of access by way of a ladder or stairway and this is provided for – but subject to the overall safety requirements set out in para 2.[39]

Where it is necessary to remove a means of protection this can be done only for the time and to the extent necessary to perform a task and once the task is complete the protection must be replaced as soon as reasonably practicable;[40] and while the task continues compensatory safety measures must be put in place.[41]

6.6.3 Schedule 3 Requirements for Working Platforms

6.6.3.1 Part 1 Requirements for all Working Platforms

Part 1 of Sch 3 to the 2005 Regulations deals with the condition of any supporting surface which must be of sufficient strength and of suitable composition to support the load put upon it while it is being used.[42]

There is a requirement that the structure itself will be of sufficient strength and rigidity.[43] Those qualities must be retained while the structure is being used, erected and dismantled or when,[44] and after, it is altered or modified.[45]

In respect of wheeled equipment inadvertent movement while work is being carried out at height must be prevented by the use of suitable devices. If the structure is free standing then it must be secured to the bearing or other structure or an anti-slip device, or some other means of equal effectiveness must be used.[46]

Paragraph 4 deals with the stability of the platform while it is being worked upon and repeats the duties found in para 3.

Paragraph 5 deals with safety aspects in respect of the actual work undertaken on the platform and repeats the duties under Sch 1 save in respect of moving parts.

Paragraph 5 of the schedule deal with stability of and working safely on the platform and repeat the duties in Sch 1.

---

39 SI 2005/735, Sch 2, para 5(1).
40 SI 2005/735, Sch 2, para 5(2).
41 SI 2005/735, Sch 2, para 5(3).
42 SI 2005/735, Sch 3, para 2.
43 SI 2005/735, Sch 3, para 3(a).
44 SI 2005/735, Sch 3, para 3(d).
45 SI 2005/735, Sch 3, para 3(e).
46 SI 2005/735, Sch 3, para 3(c).

Paragraph 6 prohibits the loading of a working platform in a manner that could cause a risk of collapse or a deformity that could affect its safe use.

6.6.3.2 Part 2 Additional Requirements for Scaffolds

Part 2 of Sch 3 to the 2005 Regulations provides for detailed planning by competent persons in the erection and dismantling of scaffolding.

The paragraphs make a distinction between standard configurations and more complex arrangements that dictate the degree of calculation of strength and documentation relating to planning prior to erection, alteration or dismantling.

In practical terms a scaffold collapse will likely arise as a result of some fault and it will not be difficult to establish liability. However, this part of the schedule will be more relevant if there are flaws in the design of the scaffold. In particular, parts of a scaffold may be put out of use from time to time and in such circumstance para 11 requires that it shall be clearly identified as out of use in accordance with the Health and Safety (Signs and Signals) Regulations 1996[47] to prevent access to the danger zone.

6.6.4 Schedule 4 Requirements for Collective Safe Guards for Arresting Falls

Schedule 4 to the 2005 Regulations relate to the use of nets and airbags, etc. Such devices can only be used if:

(a) a risk assessment has demonstrated that the work activity can, so far as it is reasonably practicable, be performed safely while using it and while perfecting its effectiveness;

(b) the use of other safer work equipment is not reasonably practicable and a sufficient number of the available persons have received adequate training specific to the safeguard including rescue operations;

(c) a sufficient number of persons have received adequate training specific to the safeguard, including rescue procedures.[48]

Any safeguard shall be suitable and of sufficient strength to safely arrest the fall of any person having to use it.[49] There are provisions relating to anchoring such safeguards securely and the means of attachment must be suitable and of sufficient strength and stability for safety purposes.[50] Finally, the safeguard itself should not be such that it could cause injury

---

47 SI 1996/341.

48 SI 2005/735, Sch 4, para 2(a)–(c)

49 SI 2005/735, Sch 4, para 3.

50 SI 2005/735, Sch 4, para 4.

to a person falling into it and sufficient steps must be taken to prevent a person being injured when falling onto a device.[51]

6.6.5 Schedule 5 Requirements for Personal Fall Protection Systems.

Schedule 5 to the 2005 Regulations relates to work restraints, work positioning, fall arrest and rope access.

These techniques are rarely used and will only come into play on very rare occasions. This will be the safety system of last resort and there are stringent requirements by way of a risk assessment to show that the work can be performed safely, so far as it is reasonably practicable, and the use of other safer work equipment is not reasonably practicable. The work can only be carried out by those who have adequate training in the use of the equipment including rescue procedures.[52]

The schedule goes on to provide that the protection system must be suitable and of sufficient strength that it should, where necessary, fit the user and be correctly fitted to prevent the worker falling from it.[53]

Parts 2 to 5 of the Schedule set out detailed measures to be taken where such equipment is used.

6.6.6 Schedule 7 Particulars to be Included in a Report of Inspection

Schedule 7 to the 2005 Regulations provides that the following particulars should be included in a report of an inspection required under the regulations:

(1) the name and address of the person for whom the inspection was carried out;

(2) the location of the work equipment inspected;

(3) a description of the work equipment inspected;

(4) the date and time of the inspection;

(5) details of any matter identified that could give rise to a risk to the health or safety of any person;

---

[51] SI 2005/735, Sch 4, para 5.

[52] SI 2005/735, Sch 5, para 1.

[53] SI 2005/735, Sch 5, para 12.

(6) details of any action taken as a result of any matter identified in para 5;

(7) details of any further action considered necessary;

(8) the name and position of the person making the report.

6.7 SCHEDULE 6 LADDERS

In practical terms injuries involving ladders, including stepladders, are among the most common matters that will arise in personal injury practice. This is dealt with specifically in Sch 6 to the 2005 Regulations.

6.7.1 Cases involving Defective Ladders

In terms of the legal requirements these are relatively straightforward matters to deal with. All work equipment must be of sufficient strength and suitable for the use to which it is put. If a ladder breaks then there will be a breach of an absolute duty. The practical difficulty in such a case will be that of proving causation. Sometimes the ladder in use will be missing after the accident and there may be an allegation that the claimant simply missed his footing.

6.7.2 Misuse of Ladders

The starting point is to understand that a ladder is a useful device for getting from one point to another – once that role is exceeded, by using a ladder as a workplace, or a working platform, the position becomes more dangerous. In terms of litigation the two most important parts of the schedule are the first and last paragraphs.

Under para 1 of Sch 6 to the 2005 Regulations the employer is required to carry out a risk assessment under reg 3 of the Management Regulations that demonstrates that the use of more suitable work equipment is not justified because of the low risk and

(a) the short duration of use; or

(b) the existing conditions on site which cannot be altered.

Paragraph 10 of Sch 6 to the 2005 Regulations requires:

> 'that every ladder shall be used in such a way that –
> (a) a secure handhold and secure support are always available to the user; and
> (b) the user can maintain a safe handhold when carrying a load unless, in the case of a step ladder, the maintenance of a handhold is not practicable when a load is carried, and a risk assessment under

> regulation 3 of the Management Regulations has demonstrated that the use of a stepladder is justified because of -
> (i) the low risk; and
> (ii) the short duration of use.'

The HSE recommends that only light materials or tools should be carried on a ladder, up to 10 kilograms, and that 'short duration' is taken to be between 15 and 30 minutes to do the task.

An essential concept in the safe use of ladders is the 'three points of contact' rule.[54] This means that both feet and one hand must be kept in contact with the rungs of the ladder. In addition the whole or the bulk of the body should remain within the limits of the ladder –overreaching to one side of the ladder is one of the most common accident scenarios.

If it is necessary to depart from the three points of contact then consideration should be towards doing the job from a different working platform or taking steps to prevent or minimise the risk and or consequences of the fall.

Using step ladders is different because often they will be used at moderate heights and both hands are often used, e.g. to change a light bulb or hang paper. The HSE advises that for a typical step ladder the three prime rules are:

- The top of the ladder must face towards the work activity. If the ladder is placed side on then there is a much greater risk of the ladder being displaced.

- The body should not lean away from the sides of the step ladder.

- There should be at least three unused steps above the feet of the user.

The remainder of Sch 6 sets out commonsense propositions in relation to the use of ladders; paras 2–9 are set out below:

> '**2.** Any surface upon which a ladder rests shall be stable, firm, of sufficient strength and of suitable composition safely to support the ladder so that its rungs or steps remain horizontal, and any loading intended to be placed on it.
> **3.** A ladder shall be so positioned as to ensure its stability during use.
> **4.** A suspended ladder shall be attached in a secure manner and so that, with the exception of a flexible ladder, it cannot be displaced and swinging is prevented.
> **5.** A portable ladder shall be prevented from slipping during use by -
> (a) securing the stiles at or near their upper or lower ends;
> (b) an effective anti-slip or other effective stability device; or
> (c) any other arrangement of equivalent effectiveness.

[54] See the HSE website www.hse.gov.uk/falls/ladders.htm.

6. A ladder used for access shall be long enough to protrude sufficiently above the place of landing to which it provides access, unless other measures have been taken to ensure a firm handhold.

7. No interlocking or extension ladder shall be used unless its sections are prevented from moving relative to each other while in use.

8. A mobile ladder shall be prevented from moving before it is stepped on.

9. Where a ladder or run of ladders rises a vertical distance of 9 metres or more above its base, there shall, where reasonably practicable, be provided at suitable intervals sufficient safe landing areas or rest platforms.'

6.7.3 Commentary

It is likely that the intention of para 5 of Sch 6 to the 2005 Regulations is to remove the notion of manual footing of a ladder found in the old regulatory regime. In this respect the HSE website contains no illustration of the practice but does contain many illustrations of mechanical devices that can be used to secure the foot of the ladder.

In respect of para 6 the HSE recommends that an access ladder should extend one metre above the landing area.

Paragraph 9, read in isolation, might indicate that a run of ladders in excess of 9 metres may be permissible without landing areas. However, the employer would have would have to justify its use by way of a risk assessment under para 1; this would lead to consideration of not using a ladder at all above that height. That might well prove very difficult because the use of a longer run of ladders is unlikely to be an activity which could be described as one carrying a low risk of injury.

One matter, surprisingly, that is not dealt with in the regulatory scheme is the pitch of the ladder. The HSE Guidance states that the ladder should be pitched at a ratio of 1:4; where 4 is the height of the resting place of the ladder and 1 is the distance of the base of the ladder from the bottom of the wall against which the ladder is to rest.

Regulation 7 sets out the requirements in respect of the integrity of any ladder – there is a strict duty of suitability for purpose. The HSE recommends Class 1 'Industrial'[55] or EN131[56] ladders or stepladders for use at work.

---

[55] BS 1129: 1990 *British standard specification for portable timber ladders, steps, trestles and lightweight stagings* British Standards Institution and BS 2037: 1994 *Specification for portable aluminium ladders, steps, trestles and lightweight stagings* British Standards Institution.

[56] BS EN 131-1 *Ladders.* Part 1: *Specification for terms, types, functional sizes* (1993) and BS EN 131-2 *Ladders.* Part 2: *Specification for requirements, testing, marking* (1993) British Standards Institution.

6.8 PRACTICAL MATTERS

These regulations should improve the prospects of success for an employee who falls off a ladder and it should in all but the rarest of cases avoid a finding of no liability as arose in the county court case of *Sharp v Elnaugh and Sons Ltd.*[57] The claimant, an electrician had to attend alone at a church to deal with a problem relating to three overhead heaters that were not working. The claimant carried a step ladder on his van that the employer knew that he had; but it was only five rungs tall. He had to borrow a ladder at the church and got assistance from a steward who footed the ladder. There came a point when the footer was asked by the claimant to turn off the electricity and while he was gone the claimant went back up the ladder. The ladder slipped and he fell to the ground.

The judge found that there was no breach of duty. The basis of the decision in this case was that the claimant was an experienced man and that the employer did not need to carry out a risk assessment or do anything more than make available a health and safety handbook (which the claimant had never read). The judge found that if the claimant had needed help he could have asked for it and it would have been provided. The judgment also refers to the lack of a risk assessment under the Management Regulations (at that time breach did not give rise to civil liability). The judge took the view that the failure to make an assessment might be negligent at common law but that an employer was able to look to the employee to use his initiative and common sense. When confronted with the House of Lords case of *Boyle v Kodak*,[58] the judge went on to find it irrelevant as, in his opinion, there had been no breach of statutory duty by the employer. The Recorder said:

> 'This case is authority for the principle that, where a breach of statutory duty on the part of an employer has been proved, the claimant is entitled to some damages, even where he is also negligent, unless it can be shown that the employer's breach did not in fact contribute to the accident.'

Under the Work at Height Regulations 2005 this case would now be decided differently, although many observers believe that it should have probably have been decided differently in any event.

There is the obvious point that footing of ladders is no longer an acceptable practice. There is now an absolute obligation to carry out a risk assessment. Therefore, Mr Sharpe would have been told how to do the job or at least how to work at height. The employer would have had to justify the use of a ladder as a low risk way of doing the job and that it was not reasonably practicable to do it some safer way. The case does not mention any training in working at height which is again now a mandatory requirement; indeed, it appears that the claimant was allowed

[57] (unreported) 4 December 2006, Colchester County Court.
[58] [1969] 1 WLR 661.

to work without having read the handbook – let alone receive actual training in working at height. The claimant was cross-examined and admitted that reading the policy would not have made any difference – ie he seems to have accepted that he ought to have known better; re-examination on this point might have been useful. However, this misses the point that training and repeated refreshers are there to improve knowledge and awareness and to prevent accidents or at least to reduce the risk to the lowest level reasonably practicable.[59] If Mr Sharp had been trained to assess the risks in accordance with the regulations (even those in force at the time) it is arguable that he would not have used a ladder at all or been more cautious.

The approach of the court in *Sharp* can usefully be contrasted with the High Court judgment in *Mason v Satelcom.*[60] There the claimant, again working off his employer's premises, had to access a box containing circuits etc some 8 feet off the ground used a 5-foot ladder that he found near the box. The ladder was angled against the wall and the claimant had to reach up to and into the box to remove a component. The rungs of the ladder were rounded and, inevitably, he fell off.

The judge, after hearing argument concerning the claimant using his imitative and having some 15 years' experience, said (para 52):

> 'In my judgment, in the present case the breaches of statutory duty under the Construction Regulations 1996 and the Work Equipment Regulations 1998, which I have found established against the Defendants, were plainly causative of the accident. Those Regulations fairly and squarely placed upon the Defendants the onus of providing safe means of access to the Claimant's place of work and there being suitable work equipment for his use. The fact that the Defendants chose to leave the Claimant to use his initiative to select his means of access to equipment he had to work upon and to choose or obtain work equipment for himself cannot, in my judgment, absolve the Defendants of the consequences of their breaches of the Regulations and cannot provide any basis for concluding that those breaches were not causative of the Claimant's accident.'

There was judgment for the claimant but also a significant finding of contributory negligence of one-third.

The introduction of the mandatory risk assessment and the requirement for competency arguably increase the weight of the duty on the employer. This will not extinguish arguments of contributory negligence but it should have an impact on the extent to which judges find an employee guilty of it.

---

59 See *Walsh v TNT UK Ltd* 2006 CSOH 149 and *O'Neil v DSG Retail Ltd* [2002] EWCA Civ 1139.

60 [2007] All ER (D) 377 (Jul), QB.

CHAPTER 7

WORK EQUIPMENT

7.1 INTRODUCTION

In Chapter 1, attention is given to the general principles that underpin the law of negligence and, specifically, to the employer's duty of care arising at common law. In much the same way as for manual handling and workplace safety, 1 January 1993 represented a watershed when considering an employer's responsibilities in the context of equipment.

The above date marked the implementation of the Work Equipment Directive 89/655/EEC through the Provision and Use of Work Equipment Regulations 1992[1] ('the 1992 Regulations'). The Work Equipment Directive was amended by Directive 1995/63/EC, which in turn was implemented in the United Kingdom through the Provision and Use of Work Equipment Regulations 1998[2] ('PUWER'). These rules revoke and replace the 1992 Regulations and came into effect on 5 December 1998.[3]

PUWER remains the principal statutory instrument governing the safety of equipment in an employment context. It is supplemented by a number of more closely defined regulations that regulate the provision and use of specific types of work equipment.

In Chapter 8 below, individual consideration is given to among the most far-reaching of these rules, namely the Personal Protective Equipment at Work Regulations 1992,[4] the Health and Safety (Display Screen Equipment) Regulations 1992[5] and the Lifting Operations and Lifting Equipment Regulations 1998.[6]

The above enactments are wide-ranging and strict in their operation, offering broad protection against the risks inherent in the everyday use of

1 SI 1992/2392.
2 SI 1998/2306.
3 Note that, by virtue of reg 37, the operation of regs 25–30 (Mobile Work Equipment) was deferred until 5 December 2002 for equipment already in use upon entry into force of PUWER.
4 SI 1992/2966.
5 SI 1992/2792.
6 SI 1998/2307.

the work equipment. In consequence, the common law is now of very limited relevance in the context of equipment safety.[7]

PUWER, as with other regulations, is supplemented by an approved code of practice ('ACOP') published with accompanying guidance by the HSC.[8] The ACOP, as discrete from the Guidance, is admissible pursuant to statute within criminal proceedings. The ACOP and Guidance are generally considered to be highly relevant in the context of civil claims as illustrating good practice.[9]

7.2 PUWER – AN OVERVIEW

In general terms, PUWER prescribes that work equipment must be:

- suitable for purpose;
- safe for use, maintained in a safe condition and, where appropriate, inspected to ensure that it remains so;
- used only by people who have received the necessary information, instruction and training to allow them to do so safely, i e without risk to themselves or others; and
- accompanied by any appropriate safety measures, e g protective devices, markings and warnings.

The regulations are divided into four main parts:

- *Introduction* – directed to issues of interpretation and application;
- *General* – setting out the regulations that are of generic application to all work equipment;
- *Mobile work equipment* – setting out supplementary requirements; and
- *Power presses* – setting out the supplementary requirements.

[7] For detailed commentary on the pre-1993 case law, see Munkman *Employer's Liability at Common Law* (11th revised edn).

[8] *Safe use of work equipment: Provision and Use of Work Equipment Regulations 1998* (L24) approved and issued pursuant to s 16(1) of the HSW 1974.

[9] See dicta of Smith LJ in *Ellis v Bristol City Council* [2007] EWCA Civ 685 (CA) at [32]–[33].

7.3 INTERPRETATION OF PUWER

Further to the guidance offered in Chapter 2 above, two central principles must be observed when interpreting the meaning and effect of PUWER in context. Firstly, the legislation was enacted for the protection of employees and each provision is to be read so as to give effect to this objective to the extent reasonably permitted by its wording.

Secondly, PUWER was intended to give effect to the Work Equipment Directive and its language must be construed so far as is possible to achieve the result intended by the Directive.

Pursuant to PUWER reg 2(1), 'Work equipment' is defined as meaning 'any machinery, appliance, apparatus, tool or installation for use at work (whether exclusively or not)'. The term, as defined, is intended to be extremely wide in its scope. It covers almost anything used at work and in practice it is more readily defined by exclusion.

For example, livestock does not constitute work equipment and privately owned vehicles also fall outside the ambit of PUWER. Substances also fall outside the definition, for example a gas, liquid or mixture stored or processed within a workplace, though direct[10] and indirect[11] protection may be offered elsewhere.

Confusion may arise where it is unclear whether an offending article or appliance constitutes work equipment or part of a workplace (including its equipment, devices or systems) to which the Workplace (Health, Safety and Welfare) Regulations 1992 apply.

Such uncertainty is illustrated by the Scottish case of *Beck v United Closures & Plastics Ltd.*[12] Lord McEwan held that two heavy doors providing access to a closed room containing fixed machinery constituted work equipment, but not a 'workplace' or 'device' for the purposes of Workplace (Health, Safety and Welfare) Regulations 1992. The judge came to this conclusion on the basis that the term 'workplace' contemplated 'things which are open places'.[13]

In *Annobil v George Payne & Co*,[14] Judge Chapman was content that failure to fit an appropriate sensor and emergency stop mechanism to a roller blind providing employee access to a confectionary factory was a

---

10 Detailed consideration of the statutory control of substances hazardous to health is beyond this work. For detailed commentary, see *Munkman on Employer's Liability* (14th edn), ch 22.

11 For example, the obligations arising under reg 12(3) of the Workplace (Health, Safety and Welfare) Regulations 1992, SI 1992/3004.

12 [2001] SLT 1299, OH.

13 See also *Lewis v Avidan Ltd* [2005] EWCA Civ 670 (CA) per May LJ at [13].

14 [2002] EWHC 1061, [2002] All ER (D) 100, QB.

breach of both reg 18(1) of the Workplace (Health, Safety and Welfare) Regulations 1992 and reg 5(1)[15] of the 1992 Regulations.

Similarly, in *Wright v Romford Blinds & Shutters Ltd*,[16] Mr Michael Harvey QC (sitting as a deputy judge of the QBD) was satisfied that the transit van from which Mr Wright fell suffering a severe head injury was 'work equipment' for the purposes of PUWER and that the vehicle's roof rack, which was being used as temporary platform, was a 'workplace' for the purposes of the Workplace (Health, Safety and Welfare) Regulations 1992.

It is submitted that the respective provisions apply co-extensively, to the extent permitted by their wording, consistent with the overall aim of providing comprehensive protection. In any case of doubt, relevant allegations under both PUWER and the Workplace (Health, Safety and Welfare) Regulations 1992 should be made. There should be few if any cases where, if properly applied, PUWER and/or the Workplace (Health, Safety and Welfare) Regulations 1992 will not provide an appropriate civil remedy when, for example, considering issues of suitability or maintenance.

When identifying 'work equipment' for the purposes of PUWER, particular caution should be exercised when third party reliance is placed upon the case of *Hammond v Metropolitan Police Comr.*[17]

Mr Hammond worked as the police equivalent of a breakdown and recovery driver, specifically under the employ of Metropolitan Police Commissioner. he was required to perform maintenance work on a van operated by the second defendant, the Metropolitan Police Authority. In order to deal with what he perceived to be the vehicle's problem, Mr Hammond attempted to remove one of the wheel bolts on the front offside wheel with a knuckle bar and socket. The wheel bolt sheared unexpectedly, causing Mr Hammond to fall awkwardly, and he thereby suffered injury.

The trial judge considered himself bound to follow the approach of Lord Abernethy in the Scottish case of *Kelly v First Engineering Ltd.*[18] Mr Kelly was a railway track engineer who, in not dissimilar circumstances to Mr Hammond, injured himself whilst attempting to remove a bolt securing a 'fish plate' – a flat piece of metal joining one rail to another.

---

15 Now reg 4(1) of PUWER.

16 [2003] EWHC 1165, QB.

17 [2004] EWCA Civ 830, [2004] ICR 1467 (CA). For detailed commentary upon this case, see Tomkins 'The antidote to '*Hammond*'?' (2006) 2 JPIL 112-118 and *Munkman on Employer's Liability* (14th edn), paras 20.10-20.15.

18 [1999] SCLR 1025, OH.

In *Kelly*, argued with specific reference to reg 2(1) of the 1992 Regulations, the court held that the bolt constituted work equipment.

The trial judge's approach accorded with commentary suggesting that the term 'work equipment' might be construed consistently with case law decided under the Employers' Liability (Defective Equipment) Act 1969 ('the 1969 Act').

In *Hammond*, however, May LJ held that the 1992 Regulations were instead concerned exclusively with what might loosely be described as 'tools of the trade', ie things that an employee works with, rather than things being worked upon, and he thus allowed the defendant's appeal.

This ruling appears to contradict the plain wording of reg 2(1) of the 1992 Regulations, which defined 'use' by reference to 'repairing, modifying, maintaining, servicing and cleaning ...'.

Moreover, the House of Lords had expressly rejected the drawing of an artificial distinction between 'tools of the trade' and articles or materials provided by employer in the course of business in *Knowles v Liverpool City Council.*[19] Here, the House of Lords confirmed that a flagstone constituted 'equipment' for the purposes of the 1969 Act.

The 1969 Act had been introduced to abrogate the effect of *Davie v New Merton Board Mills Ltd*[20] by which, at common law, an employee would have no remedy for injury or loss caused by a failure of equipment supplied by his employer, where the latter had purchased the article from a reputable supplier and the relevant defect was hidden. It is noteworthy, that *Knowles* does not appear to have been considered by the Court of Appeal in *Hammond*, and it is at least questionable whether the same result would have been achieved had this authority been taken into account.[21]

May LJ, with whom Buxton LJ agreed, also considered in obiter that a valid distinction might be drawn between work undertaken by an employee upon equipment belonging to his employer and equipment belonging to a third party when considering the application of the 1992 Regulations.

---

19 [1993] 4 All ER 321, HL.

20 [1959] 1 All ER 346, HL.

21 In particular, when providing the lead judgment in *Knowles*, Lord Jauncey of Tullichettle made the following, prescient observation: 'To give one example which I put in argument to counsel, a pump manufacturer buys in tools required for assembling the pumps as well as some components including the bolts for holding together the two parts of the housing. Workman A is tightening a bolt which sheers and injures his eye. Workman B is tightening a similar bolt but his spanner snaps causing him a similar injury. If the appellants are right, workman B could proceed under s 1(1) of the 1969 Act, but workman A would have no remedy thereunder. My Lords, I cannot believe that Parliament can have intended the Act to produce results such as these'.

Whether or not this distinction might have been sustainable, it is no longer relevant under PUWER, as amended. The decisive factor is now whether the relevant article had been 'provided for use or used' as work equipment, not whether the equipment had been provided as between the relevant employer and the injured party.

For example, the van being serviced by Mr Hammond was clearly provided as work equipment to the relevant police driver and would now be caught by the express wording of Regulation 3(2) of PUWER, as absent from the 1992 Regulations.

Thus, in *Ball v Street*,[22] the Court of Appeal held that strict liability attached when the claimant was injured whilst in the process of adjusting agricultural machinery belonging to a neighbouring farmer.

In any case regarding doubt as to the application of PUWER in this context, it would be prudent to also place direct reliance on the 1969 Act, which provides that an employer will be deemed liable for any defect in equipment supplied by a third party, where personal injury results to the employee from use of such equipment in the course of his or her employment. Further to *Knowles* above, the 1969 Act can properly be said to apply to almost any article supplied by an employer for use in connection with the activities of the business and there are very few, if any, items of work equipment that are not sourced in this way.

The continuing relevance of the 1969 Act is illustrated by the recent Scottish case of *Stephen v Peter*.[23] Mr Stephen was injured in a road traffic accident, whilst driving his employer's articulated lorry. He went round a bend too fast and the lorry toppled over onto its side. As a result of the accident, Mr Stephen suffered injuries of the utmost severity.

Mr Stephen established that the tachograph and speed limiter fitted to the lorry had been wrongly calibrated, so that instead of preventing the vehicle from exceeding 56mph, it allowed the vehicle to be driven at speeds of up to 62mph. The lorry would not have overturned with a secure load at the bend in question if it had been driven at a speed of 56mph or less and the employer was thus held liable under s 1 of the 1969 Act, concurrently with the obligations arising under reg 6(1) of PUWER 1992.

Whilst satisfied that Mr Stephen had been driving the lorry at an unsafe speed and that his actions were negligent, the trial judge was not persuaded that such conduct amounted to a new intervening act so as to break the chain of causation. It was a clearly a recognised and foreseeable risk that lorry drivers might drive their vehicles too fast.

---

22 [2005] EWCA Civ 76, CA.

23 [2005] SCLR 513, [2005] Rep LR 53, OH.

The phrase 'for use at work' is also a new addition to PUWER. It is noted, at para 67 of the ACOP, that this phrase is defined by s 52(1) of the Health and Safety at Work etc Act 1974 in the following terms:

> 'an employee is at work throughout the time he is in the course of his employment, but not otherwise and a self-employed person is at work throughout such time as he devotes to work at a self-employed person.'

In *PRP Architects v Reid*,[24] Pill LJ preferred to regard the issue of whether an employee was acting 'in the course of his employment' as important, but not a decisive factor in this context. The Court of Appeal in this case held that use of a communal lift by an employee at the end of her working day constituted use of work equipment 'at work' for the purposes of reg 3(2) of PUWER, as suggested at para 68 of the ACOP.

Similarly, a resident offshore oil rig worker was using a suspended, access ladder to descend from a bunk bed whilst 'at work', having been asleep during a rest period, when it gave way causing him to fall and suffer injury.[25]

7.4 APPLICATION OF PUWER

To whom then, do the requirements of PUWER apply? The Regulations now apply to three groups:

- an employer in respect of work equipment provided for use or used by his employee – reg 3(2);
- a self-employed person, in respect of work equipment used by him at work – reg 3(3)(a);
- anyone exercising control of (i) work equipment, (ii) anyone using, supervising or managing the use of work equipment or (iii) the way in which work equipment is used at work – reg 3(3)(b).

First and foremost, the relegations are applicable to employers in respect of all work equipment provided for use or, in fact, used by their employers at work.

It is worth remembering that PUWER does not create a discrete, stand alone duty to provide all of the equipment necessary for an employee to safely undertake a particular task.

In this wider context, it is necessary to look to the specific risk assessment obligations arising under other statutory provisions, for example the

---

24 [2006] EWCA Civ 1119 (CA) at [21]-[23].

25 *Robb v Salamis (M & I) Ltd* [2005] SLT 523; reversed upon separate grounds [2007] 2 All ER 97, HL.

Manual Handling Operations Regulations 1992,[26] the Construction (Health, Safety and Welfare) Regulations 1996[27] and in particular the Management of Health and Safety at Work Regulations 1999.[28]

The case of *Allison v London Underground Ltd*[29] emphasises the correct approach to the question of 'What the employer ought to have known about the risks inherent in his own operations?' and thus, for example, what matters should be taken into account when selecting and implementing safe use of equipment. In the context of an alleged breach of training under reg 9, Smith LJ observed as follows:[30]

> '...what he ought to have known is (or should be) closely linked with the risk assessment which he is obliged to carry out under Regulation 3 of the 1999 Regulations. That requires the employer to carry out a suitable and sufficient risk assessment for the purposes of identifying the measures he needs to take to comply with the requirements and prohibitions imposed upon him by or under the relevant statutory provisions. What the employer *ought* to have known will be what he *would* have known if he had carried out a suitable and sufficient risk assessment. Plainly, a suitable and sufficient risk assessment will identify those risks in respect of which the employee needs training. Such a risk assessment will provide the basis not only for the training which the employer must give but also for other aspects of his duty, such as, for example, whether the place of work is safe or whether work equipment is suitable ....'

Necessity is aptly described as the 'mother of invention'. Where an employee is left to make use of whatever lies to hand to complete a task, it is no defence for an employer to suggest that use by an employee of unsuitable equipment for example, thereby creating a risk of injury, does not invoke PUWER simply because the article belongs to or was supplied a third party.

In *Barnett v Scottish Power*,[31] the Court of Appeal was invited to consider the position of a part-time, electricity meter reader. In the ordinary course of her employment, Mrs Barnett was dispatched to take a reading at domestic premises. The meter was positioned at significant height and it was necessary for her to use a small, collapsible wooden chair to reach the same. The chair was provided at Mrs Barnett's request by the householder.

Whilst there was some ambiguity as to the precise mechanism, her fall from the chair was attributed to negligence on the part of her employer, amongst other matters, in failing to provide her with suitable work equipment. The Court of Appeal expressly declined to make any finding

---

26 SI 1992/2793.
27 SI 1996/1592.
28 SI 1999/3242.
29 [2008] EWCA Civ 71, CA.
30 [2008] EWCA Civ 71, CA at [57].
31 [2002] EWCA Civ 104, CA.

as to the applicability of the 1992 Regulations in this context having been satisfied that no appeal lay from the trial judge's findings at common law.

It is worth noting that Judge LJ expressed his particular dissatisfaction with the defendant's plea of contributory negligence in this matter. Mrs Barnett had made repeated requests of her employer to be provided with proper equipment and then attracted criticism having suffered injury whilst doing the best she could to get on with her job. It was held that the defendant's plea of contributory negligence was unsound in principle against this background.

In *Mason v Satelcom Ltd*,[32] the High Court was concerned with the case of an IT field service engineer who was dispatched to fix a computer fault at the premises of a third party customer.

The fault was located in a server cabinet secured on a wall at a height of about 8ft. The employer had failed to provide any suitable access equipment and Mr Mason thus used a wooden ladder he found propped up against the wall of the room in which the server cabinet was located. The ladder proved too short to safely perform the task and, whilst overstretching, Mr Mason lost his balance and fell, suffering injury.

Having found that Mr Mason's employer had breached reg 5(1) of the Construction (Health, Safety and Welfare) Regulations 1996 in failing to provide him with suitable and sufficient safe access to the server cabinet as a place of work, His Honour Judge Reddihough (sitting as a deputy judge of the QBD) robustly rejected the employer's contention that PUWER did not apply in these circumstances:

> '…it appeared to be argued on behalf of the Defendants that because they did not know the Claimant was going to use this particular ladder or of its unsuitability, they could not reasonably foresee the Claimant's safety would be affected. In my judgment, such argument is misconceived because under reg 3(2) the applicability of Regulation 4 bites upon the employers simply by virtue of the Claimant's use of the ladder. In my judgment, if an employer puts its employee in a position where he has to select the equipment he uses or he uses equipment provided by a third party or left in position by a third party, and, objectively, that equipment is unsuitable within the meaning of reg 4(1) and reg 4(4)(a), then the employer is in breach of the requirements of reg 4(1). This would be particularly so where, as here, as I have found, the Defendants knew that the Claimant would have to gain access to a cabinet which was at a height of 8 feet. Accordingly, I find that the Defendants were in breach of reg 4(1) of the Work Equipment Regulations, 1998….'

Whilst this particular issue is likely to fall to the Court of Appeal for further consideration, it is submitted that PUWER can properly be said

---

32 [2007] EWHC 2540, QB.

to apply to an employee's use of work equipment belonging to a third party, wherever it is reasonably foreseeable that an employee might need, or might be tempted, to do so.

In the absence of an appropriate risk assessment, it is likely that the court will be much more ready to hold that PUWER is applicable in this context.

In circumstances where an employer has appropriately assessed the risks associated with a particular task, provided suitable work equipment and advice, instruction and/or training as appropriate, an employee's subsequent use of alternate, unsuitable work equipment is unlikely to give rise to civil liability under PUWER.[33]

A self-employed person is most likely to attract liability when another person is injured in connection with his use of work equipment. Regulation 3(3)(a) can also form the basis for an apportionment of responsibility, as illustrated by the fairly unusual facts in *Ball v Street,*[34] where the control of work equipment is shared with another.

Mr Ball, a farmer, made arrangements with his neighbour, Mr Street, for hire of the latter's services, including use of his haybob machine; a piece of farm machinery towed behind a tractor for the dual purpose of turning and scattering new mown hay and organising dried hay into rows. The machine performs these functions using rotating tines, which can be adjusted between two settings according to the task in hand. Each tine is held in place by a stiff helical coil spring. Whilst Mr Ball was adjusting the machine between operations, one of the above springs fractured without warning and was ejected causing Mr Ball to suffer an eye injury.

The court held that whilst Mr Ball had been using the haybob on his own on a Sunday, Mr Street retained control of the equipment on the basis that he specifically authorised who was to use it, where and when it was to be used and what it was to be used for. Mr Street thereby attracted liability pursuant to reg 3(3)(b).

It was accepted that the regulations operated co-extensively between Mr Ball and Mr Street, whilst the former was using the haybob. Potter LJ observed that the only way of avoiding or minimising the risk of failure was through regular maintenance by the owner of the machine, in whose overall control it remained. Against this background, any breach by

---

33 A court is more likely to find that the employee's selection of unsuitable work equipment against this background is the sole cause of any resulting accident. See, for example, the case of *Wallis v Balfour Beatty Rail Maintenance Ltd* [2003] EWCA Civ 72, CA, decided with reference to reg 17(2) of the Workplace (Health, Safety and Welfare) Regulations 1992, SI 1992/3004.

34 [2005] EWCA Civ 76, [2005] PIQR P22, CA.

Mr Ball was of the most technical and transient nature. Accordingly, Mr Ball's award was only reduced to the extent conceded in argument.

In *Mason* above, His Honour Judge Reddihough was also content that the occupiers of a closed room in which a ladder had been stored shared responsibility with the injured party's employer, under reg 3(3)(b), when it was later used for an unsuitable purpose. The occupiers knew that the ladder had been used or might be used to gain access to the server cabinet, thereby exposing the user to a risk of falling. They could have easily removed or placed a warning notice on the ladder so as negate this risk or, at least, their responsibility.

7.5 SUITABILITY OF WORK EQUIPMENT

Of the general provisions contained within PUWER, the interpretation and proper application of reg 4 have caused the most difficulties in practice.

This is perhaps unsurprising as 'suitability' is one of few abstract expressions used within the legislation. It is thus imperative for advisers to have 'context' at the forefront of their minds when assessing the potential relevance of reg 4 to an accident at work.

Regulation 4 imposes three key and comprehensively delineated obligations upon employers:

- to ensure that work equipment is so constructed or adapted as to be suitable for the purpose for which it is used or provided – reg 4(1);

- when selecting work equipment, to have regard to the working conditions and to the risks to the health and safety of persons that exist in the premises or undertaking in which that work equipment is to be used and any additional risk posed by the use of that work equipment – reg 4(2);

- to ensure that work equipment is used only for operations for which, and under conditions for which, it is suitable – reg 4(3).

Regulation 4(4) meanwhile provides that 'suitable' in this context 'means suitable in any respect which it is reasonably foreseeable will affect the health or safety of any person'.

Further to the guidance provided within the ACOP, employers are thus expected to consider the suitability of work equipment with three main factors in mind:

- the initial integrity of equipment provided;

- the places in which equipment will be used; and
- the purposes for which equipment will be required.

In a wider context, the courts are obliged to have regard to the provisions of the Framework Direct and the Work Equipment Directive when construing all of the individual provisions PUWER. There is, however, a particularly close relationship between the requirements imposed by reg 4(2) and the employer's all-encompassing risk assessment obligations under reg 3 of the Management of Health and Safety at Work Regulations 1999, as shown by the developing case law.[35]

The aim of reg 4 is to ensure that all work equipment made available to the employee may be used without creating personal risk to the user's health and safety or risk to those working nearby. The duties imposed are strict and continuing, and extend to all related aspects of an employee's work.

The requirements and their scope are best illustrated by the Scottish case of *Robb v Salamis (M&I) Ltd*[36] Mr Robb was working on the fitting out of an offshore platform in the Moray Firth. He was provided with accommodation on the platform, incidental to the carrying out of his day-to-day activities over an extended period of time, including the use of a two-tier bunk bed. The top tier had been allocated to Mr Robb and this was accessed by means of a suspended ladder, to which the employer accepted that PUWER applied.

It appeared that the ladder had been disturbed or improperly replaced by a work colleague during Mr Robb's rest period, nobody having suggested an alternative explanation. Having failed to make any check as to the ladder's position or stability, Mr Robb fell to the floor whereupon he suffered injury. A matter of months after the above accident, the owners of the platform adapted the ladders, allowing them to be screwed permanently to the bunk beds.

At first instance, the trial judge found, among other matters, that the ladders were frequently removed and replaced, that a person replacing a ladder might fail to do so properly and that, if improperly replaced, a ladder might give way during use. Notwithstanding, the trial judge rejected Mr Robb's claim under reg 4(1) (and under reg 20) on the basis that the accident was not 'reasonably foreseeable' there being no evidence of previous accidents or complaints of such an occurrence.

[35] For illustrative purposes, a number of cases are referred to within this section. For a clear analysis of the principles underlying reg 4, one might arguably look no further than the Opinion of Lord Hope in *Robb v Salamis (M&I) Ltd* [2007] 2 All ER 97 (HL) at [1]–[36].

[36] [2007] 2 All ER 97, HL.

In any event, the trial judge concluded that Mr Robb was entirely to blame for his accident having failed to check the ladder before descending the same.

Upon first-tier appeal, the Extra Division altered the trial judge's finding as to contributory negligence, reducing this to 50%, but nonetheless held that the employer had not been in breach of PUWER, somehow managing to adopt an even narrower focus of enquiry concerning the issue of foresight.

Providing the lead opinion of the House of Lords, Lord Hope roundly rejected the approach adopted by the lower courts:[37]

> '...the question of foreseeability has to be examined in its context. The aim in [regs 4 and 20] is the same. It is to ensure that work equipment which is made available to workers may be used by them without impairment to their safety or health...It is in that context that the issue of foreseeability becomes relevant. The obligation is to anticipate situations which may give rise to accidents. The employer is not permitted to wait for them to happen...
>
> ...Regulation 4(2) serves to underline this approach...It requires that an assessment of risk be carried out before the work equipment is used by or provided for persons whose health or safety may be at risk. The aim is to identify the risks to the health and safety of workers if things go wrong. It is a short and simple step, for example, to appreciate that if a ladder becomes unstable or slips while it is being used the worker is likely to be injured. That is the risk that must be faced wherever and whenever a ladder is provided. The risk of injury if such events occur is reasonably foreseeable. Work equipment that is not so constructed or adapted as to eliminate that risk is not suitable within the meaning of reg 4(1).'

Lord Hope also rejected the suggestion that equipment could necessarily be regarded as being 'suitable' on the assumption that it would be operated as intended, ie by an employee with all necessary information, instruction and training whilst exercising proper care for the task in hand:[38]

> '...When an employer is assessing the risks to which his employees may be exposed when using equipment that he provides for them to work with, he must consider not only the skilled and careful man who never relaxes his vigilance. He must take into consideration "the contingency of carelessness on the part of the workman in charge of it, and the frequency with which that contingency is likely to arise".'

---

[37] [2007] 2 All ER 97 (HL) at [25]-[26].

[38] [2007] 2 All ER 97 at [8] and further at [32]. Lord Roger expressed a concurring opinion on this issue at [52]. Their Lordships' opinions expressly disapproved the approach adopted by the Court of Appeal in *Griffiths v Vauxhall Motors Ltd* [2003] EWCA Civ 412, [2003] All ER (D) 167 (Mar) and considerable doubt is also cast thereby upon the later ruling in *Pennington v Surrey County Council* [2007] PIQR P11, CA, in this context.

Regulation 4(2) and (4) entail that the risks associated with the use of equipment are considered as a matter of a 'generality' taking into account all relevant factors. The employer will be judged on an objective basis by reference to all information that was known or might reasonably have been ascertained at the relevant time. Liability will not attach where matters come to light after the event, the existence or significance of which might not have been reasonably foreseen, but it is misleading to suggest that an employer is not to be judged with the benefit of 'hindsight'.[39]

The absence of any noted incident or complaint may be considered a relevant factor when assessing 'suitability' in this context, whilst having due regard to considerations of usage and the reliability of health and safety reporting within the relevant organisation. It cannot, however, be regarded as the decisive factor. Similarly, the making of an otherwise conscientious risk assessment will not preclude a finding that a reasonably foreseeable risk existed.[40]

Risk factors may include the hapless conduct of others, for example the failure to properly replace a ladder in the case of *Robb* above. It may also be necessary to consider the wilful conduct of others in some cases.

In *Yorkshire Traction Co Ltd v Searby*[41] the Court of Appeal expressly rejected the contention that reg 4 does not apply to protection against external risks. This case concerned the issue of whether a public service vehicle was suitable for use, without the protection of an available safety screen, against a background of violence towards drivers. Taking into account conflicting views within a highly unionised industry, no breach was identified upon the facts.

It is highly questionable whether the result in this case is sustainable in the light of *Robb* above. The language of reg 4 is not constrained by words such as 'so far as is reasonably practicable' and once a 'reasonable

39 Hindsight means no more than 'wisdom after the event'. It is clear following *Robb v Salamis (M&I) Ltd*, if it was not beforehand, that employers are expected to be wise 'before the event'. This approach reflects not only the aims of the Framework Directive, but also the realities of the workplace as encapsulated by Lord Oaksey in *General Cleaning Contractors Ltd v Christmas* [1953] AC 180 at 189-190, HL: 'Employers are not exempted from this duty by the fact that their men are experienced and might, if they were in the position of an employer, be able to lay down a reasonably safe system of work themselves. Workmen are not in the position of employers. Their duties are not performed in the calm atmosphere of a board room with the advice of experts. They have to make their decisions on narrow sills and other places of danger and in circumstances where the dangers are obscured by repetition.' Whether or not the alleged hazard in *Palmer v Marks & Spencer plc* [2001] EWCA Civ 1528, CA (a weather strip) might properly have been regarded as presenting a 'reasonably foreseeable' risk, the language employed by Waller LJ at [27], as adopted in *Griffiths v Vauxhall Motors Ltd* among other cases, should now be treated with caution.

40 See *Home Office v Lowles* [2004] EWCA Civ 985, CA, per Mance LJ at [13].

41 [2003] EWCA Civ 1856, (2004) 148 SJLB 61.

foreseeable' risk has or might have been identified, strict liability should attach for any resulting harm.[42] The Court of Appeal conducted a balancing exercise between the competing representations of various interested parties, rather than focusing simply upon whether a 'reasonably foreseeable' risk existed and whether the employer had a means to reduce or negate the same.

In *Horton v Taplin Contracts Ltd*,[43] the Court of Appeal were not prepared to accept that the deliberate toppling of a scaffolding tower by an enraged co-employee was 'reasonably foreseeable' so as render use of such equipment unsuitable in the absence of stabilisers or outriggers.

The Court of Appeal's reasoning in *Horton* on the issue of whether the scaffolding tower complied with PUWER is also difficult to reconcile with the dicta of Lord Hope in *Robb*:[44]

> '…The employer is liable even if he did not foresee the precise accident that happened . . ..As Lord Reid said in *Hughes v Lord Advocate* [1963] 1 All ER 705 at 708, [1963] AC 837 at 847, the fact that an accident was caused by a known source of danger but in a way that could not have been foreseen affords no defence….'

It was unsuccessfully argued in *Horton* that if stabilisation of the scaffolding tower with outriggers was 'necessary for purposes of health and safety' on the basis that the tower might topple in foreseeable circumstances, the example given being an accidental collision, then the absence of the outriggers must constitute a breach of PUWER even though the event which actually caused the tower to fall was something quite different.

The deliberate act of an injured party may introduce scope for argument that a particular breach of duty was not causative of the resulting loss[45] or that the act of a co-employee represented a new intervening act so as to break the chain of legal causation,[46] but it must be the position that equipment either complies with relevant requirements of PUWER or it does not. To suggest that equipment may be suitable in certain circumstances, but not in others, is clearly a reversal of the correct approach outlined by Lord Hope above.

The external factors (those other than the physical integrity of or purposes for which equipment might be used) to which an employer might

[42] This approach is consistent with the leading Scottish authority on this issue, *Skinner v Scottish Ambulance Service* [2004] SC 790, [2004] SLT 834, IH.

[43] [2002] EWCA Civ 1604, ICR 179, [2003] PIQR P12.

[44] [2007] 2 All ER 97 at [29].

[45] See, for example, *Wallis v Balfour Beatty Rail Maintenance Ltd* [2003] cited above.

[46] This was the alternate, arguably more sustainable basis upon which the Court of Appeal rejected the claimant's appeal in *Horton* [2002] EWCA Civ 1604, ICR 179, [2003] PIQR P12 per Bodey J at [24]-[28].

have regard are not limited to human actions, but might extend to a wide range of separate or interrelating considerations.

Environmental factors are likely to apply widely. The most pertinent question is likely to be 'whether equipment will be used indoors or outdoors?' Among other statutory provisions, the Workplace (Health, Safety and Welfare) Regulations 1992[47] provide for the regulation of a number of environmental factors, such as temperature, humidity and lighting. These may be less readily controlled in the open or at the premises of another.

For example, the footplate or steps of a vehicle used within an enclosed workplace may be safe when its use is so restricted, whereas adaptation, for instance by means of adherent surfacing or additional handholds, may be necessary if the same equipment is to be operated in an open yard or enclosure. Particular care is likely to be necessary when assessing the suitability of electrical or other powered equipment used in these circumstances.

Conversely, where equipment is to be used in an enclosed environment other factors may come to the fore. Are all emissions from work equipment safe or capable of control by adaptation or other means?[48] Can equipment be used safely within any constraints of space or proximity to others?[49] Ergonomic considerations may be particularly relevant against this background.

For example, a variety of powered tools may be suitable for outdoor use notwithstanding any gas, dust or vapour emissions, assuming appropriate provision of personal protective equipment, but wholly unsuited for use within an enclosed environment or safe only when adapted, for example by means of integral extraction.

The higher courts have considered the issue of suitability in the context of work equipment in a wide range of contexts, including the following:

- candles were not suitable to provide lighting within a tent used by a support worker assisting a handicapped person when a battery-powered lamp would involve a negligible risk by comparison;[50]

---

47 SI 1992/3004.

48 An employer's obligations pursuant to reg 4 may well overlap those arising under reg 6 of the Workplace (Health, Safety and Welfare) Regulations 1992, SI 1992/3004 and the Control of Substances Hazardous to Health Regulations 2002, SI 2002/2677 in this context.

49 Similarly, the Confined Spaces Regulations 1997, SI 1997/1713 and the Manual Handling Operations Regulation 1992, SI 1992/2793 may also be applicable in such circumstances.

50 *Fraser v Winchester HA* (2000) 55 BMLR 122, CA.

- the roof of a works van used as a temporary working platform for an operative handling bundles of metal shuttering was not suitable in the absence of guardrails to reduce the risk of falling;[51]

- a van used for transporting prisoners was not suitable in the absence of handholds where escorts might be required to move around the vehicle's interior during transit;[52]

- the cutting blade of a meat-slicing machine was not suitable in the absence of a finger guard so as to prevent accidental contact during cleaning operations;[53]

- a portable bench saw was not suitable for trimming long sections of wooden facia board in the absence of a run-off bench to support the length of such materials.[54]

Whilst the ruling of House of Lords in *Robb* should hopefully introduce greater consistency of approach when trial judges are asked to determine issues of suitability, two central questions have been expressly reserved for future consideration by the higher courts.

Firstly, upon which party does the burden of proof to establish compliance or failure to comply with the requirements of reg 4 rest?[55]

Consistent with the settled approach of the courts to questions of compliance with the Manual Handling Operations Regulations 1992 and with all other statutory provisions, within which the phrase 'reasonable practicability' operates as a potential defence, the duty at law and the burden of proof within any proceedings must both rest with the relevant defendant.

It would be illogical and inconsistent for an employee to whom an 'absolute and continuing duty' was owed to be placed in an inferior position both procedurally and evidentially, by comparison with a person who instead places reliance upon a qualified statutory obligation.

Secondly, is reg 4(4) an appropriate implementation of Article 5(4) of the Framework Directive?[56]

---

[51] *Wright v Romford Blinds & Shutters Ltd* [2003] EWHC 1165, QBD.
[52] *Crane v Premier Prison Services Ltd* (unreported) 2001, CL-01/3298 (unreported decision of the QBD).
[53] *English v North Lanarkshire Council* [1999] SCLR 310, OH.
[54] *Sherlock v Chester City Council* [2004] EWCA Civ 210, CA.
[55] [2007] 2 All ER 97 per Lord Clyde at [45]. Of course, if the burden of proof rests with the defendant on this issue, logically the same must follow for each provision within PUWER and for all other statutory duties in this context, except where the relevant instrument provides otherwise.
[56] [2007] 2 All ER 97 per Lord Clyde at [47].

Regulation 4(4) introduces the concept of 'reasonable foreseeability', whilst Art 5(4) speaks narrowly of,

> 'unusual and unforeseeable circumstances, beyond the employer's control, or to exceptional events, the consequences of which could not have been avoided despite the exercise of all duty care'

by way of permissible limitation. It is difficult to see how importation of a phrase that originates at common law can accurately transpose Article 5(4), the language of which appears to imply a much higher, purposive standard.

Unless and until such time as the House of Lords provide otherwise, those acting for injured parties might properly:

- invite the court to proceed on the basis that the burden of proof rests with the defendant to establish that equipment complies with the reg 4(1), once it is accepted or established by the claimant that PUWER applies to the same; and
- place direct reliance on Art 3(1) of the Work Equipment Directive, to which reg 4(1) is intended to give effect, in appropriate cases.

7.6 MAINTENANCE OF WORK EQUIPMENT

The maintenance of work equipment is regulated by regs 5, 7 and 22 of PUWER.

Further to reg 5(1), all work equipment must be maintained in 'an efficient state, in efficient working order and in good repair'. Whilst it is undoubtedly good practice to do so, there is no express requirement to keep a maintenance log. Where a maintenance log does exist, most commonly in the case of fixed machinery or larger items of powered equipment, it must be kept up to date, pursuant to reg 5(2).

The application of reg 5(1) was definitively considered by the Court of Appeal in *Stark v Post Office*.[57] Mr Stark was a postman who, whilst riding his works bicycle, suffered injury owing to mechanical failure of a brake stirrup.

It was accepted that metal fatigue or a latent manufacturing defect was the most likely cause of the stirrup's failure and that a 'perfectly rigorous examination' would not have revealed the bicycle's problem. The Court of Appeal decisively held that reg 6(1) of PUWER 1992, now reg 5(1), imposed a strict obligation to ensure that work equipment was 'at all

[57] [2000] PIQR P105.

times' maintained in an efficient state. Unexpected failure of the bicycle's stirrup amounted to a breach of this duty.

The Court of Appeal placed significant reliance upon the fact that the House of Lords had previously construed the same language in the context of the Factories Act 1937 in *Galashiels Gas Co Ltd v Millar*:[58]

> '...In the ordinary use of language one cannot be said to maintain a piece of machinery in efficient working order over a given period if, on occasion within that period, the machinery, whatever the reason, is not in efficient working order. In short, the definition describes a result to be achieved rather than the means of achieving it . . ..'

The court thus accepted that when the draftsman responsible for the wording of reg 6(1) had used language that had previously been construed authoritatively as imposing an unqualified obligation under the Factories Acts, he must be taken in the context of legislation enacted to promote the health and safety of workers to have intended a strict interpretation.

Regulation 5(1) thus provides extremely wide protection, extending to all cases where injury has been caused by the malfunctioning or failure of work equipment, irrespective of the efforts of a contentious employer to devise and implement an appropriate system of inspection, maintenance and repair.

The term 'efficient' relates to how the condition of the equipment might affect health and safety, not productivity. In *Ball v Street*,[59] it was suggested that there had not been a failure to maintain a haybob machine, a broken part from which had been ejected causing injury, when the device was still capable of being used in an otherwise effective and efficient manner. The court strongly rejected such an interpretation:[60]

> '...I do not accept the broad proposition...that, where there is an expendable part in a machine...so that the mechanism can continue working in an overall effective and efficient manner, no breach of reg 5(1) is demonstrable...The Regulation does not define the employer's duty in terms of the overall suitability of the equipment to perform the task for which it is designed. It deals with the duty to maintain it in an efficient state and working order and in good repair in respect of all of its mechanical parts so as to prevent injury to the person using the equipment...the object of the Regulations is a broad one, namely to protect workmen, and the task of the court is to view the maintenance and the condition of the machinery supplied to them from the point of view of health and safety and not that of productivity or economy ....'

---

[58] [1949] AC 275, per Lord MacDermott at 286.

[59] [2005] EWCA Civ 76, [2005] PIQR P22.

[60] [2005] PIQR P22 per Potter LJ at [44].

Ball followed *Fytche v Wincanton Logistics plc*[61] wherein the Houses of Lords had considered 'maintenance' in the specific context of reg 7(1) of the Personal Protective Equipment at Work Regulations 1992. The Court of Appeal expressly rejected the proposition that, by reference to and following *Fytche*, foreseeability was a relevant consideration when simply considering an alleged breach of reg 5(1).

Donachie v Chief Constable of Greater Manchester[62] illustrates the wide scope of reg 5(1) in another sense. Though principally concerned with issues of causation, in particular the proper application of *Page v Smith*[63] to cases involving a remote risk of psychiatric harm, *Donachie* underlines the need to think laterally about the applicability of health and safety provisions.

Mr Donachie was required to attach a tagging device to the underside of a car that the crime squad believed belonged to an underworld gang. The system operated by the team was that one officer, in this case Mr Donachie, would attach the device to the underside of the car, while the others kept watch from the relative safety of a police 'tracking' van to guard against the possibility of interruption.

Unbeknown to Mr Donachie and his colleagues, the battery cell powering the device was defective, and he had to retrieve and replace it on eight further occasions before it was successfully attached in full working order. With every approach that Mr Donachie made to the vehicle, he subjected himself to an increased risk of being caught in the act and attacked by the suspects. He became increasingly frightened, fearful of serious injury or even death if discovered, and he ultimately suffered a stress-related injury in consequence.

Whilst most cases under reg 5(1) are founded upon a direct failure of equipment, *Donachie* demonstrates that the indirect consequences of failure may equally give rise to valid claim. For example, slippages following a leakage of oil, grease or other such substance from defective machinery might equally succeed on this basis.

Regulation 7 requires employers to ensure that where the use of work equipment is likely to involve a 'specific risk', the use, repair, modification, maintenance or servicing of such equipment is restricted to those persons who have been specifically designated to perform such operations. It is further required that employers must provide adequate training to enable such tasks to be performed safely.

No definition of 'specific risks' is provided within PUWER, the ACOP or the accompanying Guidance. As to the steps required in this context,

[61] [2004] UKHL 31, [2004] 4 All ER 221.
[62] [2004] EWCA Civ 405, (2004) 148 Sol Jo LB 509.
[63] [1996] AC 155.

para 167 of the ACOP provides that risk control measures should be approached by employers on a hierarchical basis. The following steps should be taken in the order given:

- eliminating the risks, or if that is not possible;
- taking 'hardware' (physical) measures to control risks such as the provision of guards, but if risks cannot be adequately controlled in this way;
- taking appropriate 'software' measures to deal with the residual risks, eg by following safe systems of work, the provision of information, instruction and training as appropriate.

Advisers must closely examine whether an employer has approached risk assessment and reduction on the correct step-by-step basis, as indicated above.

Regulation 22 (implementing Art3 and Annex 1, para 2.13, of the Work Equipment Directive) provides that employers must take appropriate measures to ensure that all work equipment is constructed or adapted so as to allow maintenance operations to be carried out without exposing the relevant operative to a health and safety risk. This wider aspect of the duty is not qualified.

The specific duty arising under reg 22, to ensure that equipment is constructed or adapted so as to allow maintenance operations involving a health and safety risk to be carried out whilst the equipment is shut down, is qualified by the words 'so far as is reasonably practicable'.

If this is not reasonably practicable, work must be carried out in such a way as to prevent exposure to risk and appropriate measures taken to protect those carrying out such maintenance. Appropriate measures may include, for example, features limiting the power, speed or range of movement of dangerous parts during maintenance.

7.7 INSPECTION OF WORK EQUIPMENT

PUWER reg 6 provides that all work equipment must be inspected in certain circumstances to ensure that it is, and continues to be, safe for use. Any such inspection must be carried out by a competent person (though this may be an employee if they have the necessary competence and training to perform the task) and an appropriate record kept until the next inspection.

In practice, it is highly unlikely that direct reliance would be placed upon a breach of reg 6 in the context of a personal injury action. The 'paper trail' envisaged upon an employer's proper compliance with such inspection requirement is, however, often relevant on an evidential basis.

For example, if it is not possible to inspect work equipment (eg following replacement or removal) or to replicate a particular complaint or description of malfunctioning, the court may be more or less inclined to accept the injured party's account based upon the standard of the employer's record-keeping.

7.8 INFORMATION, INSTRUCTION AND TRAINING

Managers, supervisors and users of work equipment must be provided with adequate information and, where appropriate, written instructions relating to work equipment (PUWER reg 8(1)) and adequate training (reg 9(1)).

In the context of both provisions, adequate means 'adequate for the purposes of health and safety', ie so as to enable an employee to use equipment without risk to his or her health and safety to the full extent possible by such means. The requisite level of input required to comply regs 8 and 9 will very much depend upon the particular item of work equipment, the person who will be using it and the purpose for and circumstances in which it will be used.

The employer's starting point when addressing such issues should be the general risk assessment conducted for the purposes of the Management of Health and Safety at Work Regulations 1999.

It may be necessary to obtain specialist advice and input when considering the provision of information, instruction and training for some items of work equipment (or to effectively outsource this role), particularly where external expertise has been sought concerning the issue of selection under reg 4(2) above. The employer will, however, retain legal responsibility for any failure giving rise to injury.

In almost all cases, commonsense dictates that an employee should at the very least be provided with all of the health and safety information, instruction and training recommended by the supplier or manufacturer of the relevant work equipment.

When presented with a case where there is evidence of pre- or post-accident concerns or experiences of co-employees regarding the use of specific work equipment, advisers should look closely at whether and, if so, to what extent, the employer had complied with the strict duties arising under the Health and Safety (Consultation with Employees) Regulations 1996. Whilst not conferring civil liability directly, these Regulations provide that an employer must consult with employees about health and safety at work, in particular, upon:

- any change that may substantially affect their health and safety at work, for example in procedures, equipment or ways of working;
- the employer's arrangements for getting competent people to help him or her satisfy health and safety laws;
- the information that employees must be given on the likely risks and dangers arising from their work, measures to reduce or get rid of these risks and what they should do if they have to deal with a risk or danger;
- the planning of health and safety training; and
- the health and safety consequences of introducing new technology.

The employer's assessment process must have regard to all risks that may arise during use of work equipment and any precautions to be taken. The requisite level of information, instruction and training will depend upon all the circumstances of use, including the skill and competence of the relevant employee. Other regulations may contain specific training requirements. In all cases, particular attention must be paid to the training and supervision of young persons and, in appropriate cases, to the activities of experienced operatives.

Regulations 8 and 9 are both strict in their application. Whilst on-the-job or supervised training may be necessary or desirable, it is no answer for an employer to suggest, for example, that training had been placed 'in hand', but had not yet commenced or been completed. Many accidents occur within a short period of a young or inexperienced worker being asked to use unfamiliar work equipment. An employee must not be placed in a position of risk to his or her safety prior to being provided with the requisite information, instruction and training to minimise the same.

In common with the suitability and maintenance obligations, the employer's duty to provide adequate information, instruction and training relating to work equipment is an absolute and continuing one. It is no answer for an employer to suggest that training or instruction was impracticable.[64]

Similarly, where the duty to provide information, instruction and training has not been discharged, an employer is highly unlikely to succeed in arguing that the relevant employee would have disregarded adequate provision come what may.[65]

---

[64] *Pennington v Surrey County Council* [2006] EWCA Civ 1493, [2007] PIQR P11 (CA) per Pill LJ at [38]–[39] and per Arden LJ at [48]–[50].

[65] *Sherlock v Chester City Council* [2004] EWCA Civ 201 (CA) per Latham LJ at [26]. See also *Henser-Leather v Securicor Cash Services Ltd* [2002] EWCA Civ 816 (CA) per Kennedy LJ at [23] and [24] in the context of instruction regarding the use of PPE.

The employer's duty is thus a broad one and may entail the giving of very specific information, instruction and training, which closely replicates the anticipated circumstances of use.

In *Pennington v Surrey County Council*,[66] the Court of Appeal was not satisfied that a rescue worker's training with a lighter piece of hydraulic ram equipment, in the relative calm of a controlled training environment, was adequate preparation for his use of heavier equipment, albeit of a similar nature, in an actual emergency situation:

> '...Where dangerous machinery of this kind is used in stressful situations in circumstances not covered by use demonstrated in training exercises, I would incline to the view that the training was inadequate if (as here) it consisted of nothing more specific than a general instruction that the operator of the machinery should always keep his hand out of the moving parts. That mundane if necessary advice means little in extreme situations unless it is concretised with and accompanied by some advice applicable to the actual situation with which an operator is faced. On its own, that instruction tells him nothing about how he should conduct himself ....'

In *Allison v London Underground Ltd*,[67] the Court of appeal considered training in the context of a safety device or 'dead man's' handle used by train drivers, which needed to be held in the same position for extended periods while the engine was in operation. The employer consulted senior operatives and took expert ergonomic advice regarding the original design, but failed to revisit its risk assessment following a seemingly minor modification to the shape of the handle. In fact, unless accompanied by appropriate instruction and training concerning hand positioning, the modified handle presented a foreseeable risk of injury (tenosynovitis) with extended use. The court held that, having appreciated that a comprehensive risk assessment was required in the first instance, the employer should not have put the device back into service without the benefit of further advice from a suitably qualified expert. Had further advice been sought, the risk would have been identified and the employee's appeal against the dismissal of her claim was thus allowed with reference to reg 9(1).

The Court of Appeal was undoubtedly correct to empress doubt as to whether a breach reg 4(1) was not also made out on the above facts.[68] Equipment cannot be regarded as 'suitable' on the assumption that it will only be used in accordance with information, instruction and training, if such risks can be identified and eliminated at the design stage.[69]

Responsibilities towards younger workers are likely to be of generic concern to employers.

66 [2006] EWCA Civ 1493, [2007] PIQR P11, CA.
67 [2008] EWCA Civ 71, CA.
68 [2008] EWCA Civ 71 per Smith LJ at [15]
69 *Robb v Salamis (M&I) Ltd* [2007] 2 All ER 97 at [8] and further at [32].

In *Fraser v Winchester Health Authority*,[70] the Court of Appeal considered the position of a young and inexperienced care worker who was sent, alone, ill equipped and inadequately trained, to provide social care for a seriously handicapped patient on a camping trip. Upon return to their tent after nightfall, Miss Fraser attempted to change a gas cylinder on the camping stove which she had been sent along with.

She had also been provided with candles to illuminate the tent and, whilst appreciating that a naked flame presented a combustion risk, she mistakenly believed that she would be safe to perform this task whilst leaning out of the mouth of the tent. An explosion ensued and both occupants were injured.

Although the claimant accepted knowledge of the risks of her actions, the Court of Appeal upheld the trial judge's finding that her employer's failure to provide any instruction or training concerning the above task constituted a material breach of statutory duty.

It is a specific requirement that information and instruction for these purposes must be 'readily comprehensible to those concerned'.

In *Tasci v Pekalp of London Ltd*,[71] the Court of Appeal was invited to consider the position of a Kurdish man from Turkey who had limited understanding of English. He suffered serious injury whilst using a circular saw at the defendant's woodworking premises.

Whilst the trial judge dismissed the claim, primarily on the basis that Mr Tasci had overstated his experience and had not being using the saw in the manner demonstrated to him at the time of the accident, the Court of Appeal was satisfied that the employer's enquiries had been inadequate and that there had been a material failure to provide adequate information, instruction and training:[72]

> '...if a man presents himself for employment in a woodworking establishment, having come from another reputable and well-established woodworking establishment, armed with an appropriate reference or details which will enable the new employer accurately to assess, by making inquiry if necessary, the precise extent of the applicant's previous experience and training, it may be that a new employer is justified in treating the provisions [relating to information, instruction and training] as having been complied with and requiring no further steps by him, other than a relatively cursory inspection of the applicant's method of work, in order to satisfy themselves

---

[70] (2000) 55 BMLR 122, CA.

[71] [2001] ICR 633, CA. *Tasci* actually concerned reg 13 ('training and instruction') of the Woodworking Machines Regulations 1974, SI 1974/903 which were revoked by PUWER, though the Court of Appeal's observations are equally relevant in the context of the regs 8 and 9 of PUWER.

[72] [2001] ICR 633 (CA) per Wright J at [22]–[23].

> that he is a sensible person who is putting into operation the precautions and steps about which he has been instructed…
>
> …Here, this applicant was a refugee from a remote part of Eastern Europe. Any prudent employer would have recognised the high probability that he could put little or no reliance upon any representations made by the applicant for work as to his previous experience and training, on the basis that a person seeking refugee status would be prepared to say almost anything in the desperate need to obtain employment to support himself in this country. In any event, the standards of training and education that obtain in woodworking establishments in Eastern Turkey may bear little or no relation to the standards that apply in this country. In truth, the prudent employer, faced with the problems that this applicant for work would have presented, would have treated him as a total novice and have started [from the beginning] with a training and supervision regime in order to satisfy himself that his new worker was capable of operating a circular saw safely ….'

The Court of Appeal identified a number of the key considerations for employers, when determining what an individual employee requires by way of information, instruction and training:[73]

- The employer's duty is not restricted to use of work equipment that is 'intellectually demanding' – simple tasks using simple equipment may present significant risk.

- The duty entails providing a 'comprehensive explanation' using 'ordinary language' – not only as to how to use work equipment as intended, but also as to the dangers associated with using the implement or device in any reasonably foreseeable way.

- The duty involves an appraisal of whether the employee understands the training provided and the dangers involved and, thereafter, such monitoring of his or her working practices as the circumstances dictate.

- The nature and extent of the employer's duty will depend upon a wide variety of factors, including the experience, skill and/or maturity of the relevant employee, any language or other communication issues, whether the task is unusual or repetitive and the likelihood or likely severity of harm which use of the relevant implement or device may present.

The Court of Appeal's recent observations in *Tasci* regarding 'language difficulties' echo those in *Bux v Slough Metals*,[74] a pre-1974 authority concerning instruction and supervision provided to a Pakistani man relating to use of eye protection.

---

[73] [2001] ICR 633 (CA) per Pill LJ at [38]–[44].

[74] [1974] 1 All ER 262. Stephenson LJ prophetically observed at p 273 that a worker whose

Concerned by a rise in the reported incidence of severe and fatal injury among migrant workers, the HSE and TUC issued a joint press release in 2004[75] to mark the translation of new health and safety information into 21 different languages.

Whilst employers' health and safety obligations apply equally to workers from all backgrounds and nationalities, advisers should be acutely conscious of the day-to-day reality for many migrant workers.

7.9 CONFORMITY WITH COMMUNITY REQUIREMENTS

PUWER reg 10[76] applies to all items of work equipment provided for use in the premises or undertaking of an employer for the first time after 31 December 1992.

Regulation 10(1) provides that an employer must ensure that, where an applicable Community Directive concerning the safety of products was in operation at the time of first supply or at the time of initial provision for use as at work, such equipment conforms to all 'essential requirements'.

Regulation 10(2) defines the term 'essential requirements' in the context of work equipment as the,

> 'requirements relating to its design or construction in any of the instruments listed in Schedule 1 (being instruments which give effect to Community directives concerning the safety of products).'

Not all work equipment is covered by a relevant product directive and the 'essential requirements' of any applicable statutory instrument do not operate retrospectively. On a prospective basis, however, reg 10 will have the effect of promoting harmonisation of standards concerning the initial integrity of work equipment.

PUWER, in contrast to the 1992 Regulations, no longer restricts the application of regs 11–19 and 22–29 in this context. An injured person is thus entitled to rely upon PUWER and any specific product safety instrument in a claim against his or her employer.

In most instances, non-conformity of work equipment will only serve to reinforce a claim under an alternate provision of PUWER; for example

---

English was poor might need to be instructed in his own language. This is undoubtedly an accurate, present statement of the law in all instances where information, instruction and training concerning health and safety issues are required.

[75] HSE Press Release: E170:04 – 9 December 2004.

[76] As amended by the Health and Safety (Miscellaneous Amendments) Regulations 2002, SI 2002/2174 which implements Art 4 and Annex I para 2.8 of the Work Equipment Directive (as amended).

reg 4 (as evidence that equipment was not suitable). The obligation is, however, a strict one and consideration should always be given to placing direct reliance upon reg 10 where applicable.

7.10 DANGEROUS PARTS OF MACHINERY

PUWER reg 11(1) requires an employer to ensure that access to 'dangerous parts' of a machine is prevented or, if access is needed, to ensure that the machine is stopped before any part of the operative's body reaches a 'danger zone'.

The term 'danger zone' is defined by reg 11(5) as meaning,

> 'any zone in or around machinery in which a person is exposed to a risk to health or safety from contact with a dangerous part of machinery or a rotating stock-bar.'

Whilst there are many reported cases as to the meaning of 'dangerous part' for the purposes of s 14 of the Factories Act 1961 (since repealed), it is doubtful whether the earlier approach of courts, embarking upon enquiry as to whether injury was 'reasonably foreseeable' whilst an operative was making anticipated use of equipment, is applicable in the context of reg 11 having regard to the European origins of PUWER.

In *Pennington v Surrey County Council*,[77] a 'pinch point' created during a fireman's untrained use of a hydraulic power ram whilst attempting to straighten out a lorry cab that had buckled in a road accident was held to constitute a 'dangerous part' of machinery for these purposes.

Regulation 11(2) sets out a hierarchy of preventive measures required to comply with reg 11(1). It is important to note that the this regulation was amended with effect from 17 September 2002 to say that the measures required by para (1)

> 'shall consist of—
> (a) the provision of fixed guards enclosing every dangerous part or rotating stock-bar where and to the extent that it is practicable to do so, but where or to the extent that it is not, then
> (b) the provision of other guards or protection devices where and to the extent that it is practicable to do so, but where or to the extent that it is not, then
> (c) the provision of jigs, holders, push-sticks or similar protection appliances used in conjunction with the machinery where and to the extent that it is practicable to do so, and
> (d) the provision of such information, instruction, training and supervision as is necessary.'

[77] [2006] EWCA Civ 1493, [2007] PIQR P11, CA.

This means that a defence based on provision of information, instruction, training and supervision cannot any longer be sustained on its own. The *Pennington* decision was based on the old regulation in which the letter 'd' was used instead of the word 'and'. That wording left open the potential argument that the provision of information, instruction, training and supervision could be a defence when injury occurred from contact with a 'dangerous part of machinery'.

In *McGowan V W & J R Watson Ltd*[78] McGowan cut his left middle finger on the blade of a circular saw. The judge held that:

> 'as a result of some aberration or lapse in attention, [McGowan] inadvertently let go of the wood and allowed his finger to come into contact with the blade.'

There was nothing to prevent McGowan from having access to the revolving blade. It was held that accordingly the result required by the regulation was not achieved. The judge held that the defenders were in breach of their statutory duty under reg 11(1) and but for their breach of that duty, the accident would not have occurred. An appeal proceeded and failed on contributory negligence. The finding of breach of statutory duty was not challenged and the appellate court did not suggest that it should have been.

Regulation 11(3) makes specific provision that guards and protective devices must be robust, difficult to defeat and appropriately maintained. The express requirements are that such devices shall:

- be suitable for the purpose for which they are provided;
- be of good construction, sound material and adequate strength;
- be maintained in an efficient state, in efficient working order and in good repair;
- not give rise to any increased risk to health or safety;
- not be easily bypassed or disabled;
- be situated at sufficient distance from the danger zone;
- not unduly restrict the view of the operating cycle of the machinery, where such a view is necessary; and
- be so constructed or adapted that they allow operations necessary to fit or replace parts and for maintenance work, restricting access so

[78] [2006] CSIH 62.

that it is allowed only to the area where the work is to be carried out and, if possible, without having to dismantle the guard or protection device.

7.11 EQUIPMENT CONTROLS

Regulations 14-18 of PUWER (implementing and building upon para 2 of Annex 1 of the Work Equipment Directive) make detailed provision in relation to equipment controls. In summary, equipment must be provided with:

- one or more controls for starting the equipment – reg 14(1)(a);
- and which control the speed, pressure or other operating conditions where the risks after a change is greater than or of a different nature from the risks beforehand – reg 14(1)(a);
- one or more readily accessible stop controls that will bring the equipment to a safe condition in a safe manner – reg 15(1);
- and which operate in priority to any control that starts or changes the operating conditions of the equipment – reg 15(4);
- One or more readily accessible emergency stop controls – unless by the nature of the hazard, or by the adequacy of the normal stop controls (above), this is unnecessary – reg 16(1);
- emergency stop controls must operate in priority to other stop controls – reg 16(1);
- controls should be clearly visible and identifiable (including appropriate marking) – reg 17(1);
- the position of controls should be such that the operator can establish that no person is at risk as a result of the operation of the controls – reg 17(3)(a);
- where this is not possible, safe systems of work must be established to ensure that no person is in danger as a result of equipment starting – reg 17(3)(b);
- where this is not possible, there should be an audible, visible or other suitable warning whenever work equipment is about to start – reg 17(3)(c);
- Sufficient time and means shall be given to a person to avoid any risks due to the starting or stopping of equipment – reg 17(4);

- control systems should not create any increased risk, which ensure that any faults or damage in the control systems or losses of energy supply do not result in additional or increased risk, and which do not impede the operation of any stop or emergency stop control – reg 18.

The HSE make advice to employers in this context readily available.[79] In any case where the absence or failure of control devices is a feature, it may be helpful to consider these questions on a preliminary basis:

- Whether 'hold-to-run' or two-handed controls were positioned at a safe distance from the danger area.
- Whether stop and start buttons were readily accessible.
- Whether control switches were clearly marked to indicate their use.
- Whether operating controls were designed and placed so as to avoid accidental operation, eg by shrouding starter buttons and pedals.
- Whether interlocked or trapped key systems for guards might have prevented an operator or maintenance worker from entering a danger area before a machine had stopped.
- Whether emergency stop controls were situated within easy reach of operatives, particularly on larger machines, so they could be operated quickly in an emergency.

7.12 OTHER CONSIDERATIONS

PUWER reg 12 provides that an employer must take measures to prevent an employee being exposed to any health and safety risk arising from a number of specified hazards, namely:

- any article or substance falling or being ejected from work equipment;
- rupture or disintegration of parts of work equipment;
- work equipment catching fire or overheating;
- the unintended or premature discharge of any article or of any gas, dust, liquid, vapour or other substance which, in each case, is produced, used or stored in the work equipment;

[79] For example, '*Using work equipment safely*' (INDG 229) from which the above questions have been adapted.

- the unintended or premature explosion of the work equipment or any article or substance produced, used or stored in it.

If is not reasonably practicable to prevent such risks, an employer must take adequate steps to control the same. Emphasising the pervasive hierarchical approach to preventative measures, the employer must comply with this obligation, so far as is reasonably practicable, without placing any reliance upon the supply of personal protective equipment or by providing specific information, instruction or training. The employer must look to minimise the effects of the hazard, as well as to reducing the likelihood of its occurrence.

The above requirements do not apply where a more onerous and closely defined statutory framework exists to control a specified hazard, as provided by reg 12(5):

- Ionising Radiations Regulations 1985;[80]
- Control of Substances Hazardous to Health Regulations 1994;[81]
- Control of Asbestos Regulations 2006;[82]
- Control of Noise at Work Regulations 2005;[83]
- Construction (Head Protection) Regulations 1989;[84]
- Control of Lead at Work Regulations 1998;[85]
- Control of Vibration at Work Regulations 2005.[86]

Regulation 13 provides protection against high or very low temperature. Work equipment, all parts and any article or substance produced, used or stored in work equipment must be protected, where appropriate, so as to prevent burns, scalds or sears.

The qualification 'where appropriate' must mean any instance in which an employee is at risk of 'coming into contact or coming too close' to the source of such high or very low temperature, if the provision is to be interpreted consistently with the Work Equipment Directive.[87]

---

80 SI 1985/1333.
81 SI 1994/1657.
82 SI 2006/2739.
83 SI 2005/1643.
84 SI 1989/2209.
85 SI 1998/543.
86 SI 2005/1093.
87 See Work Equipment Directive, Annex I at para 2.10.

Regulation 18 provides that, where appropriate, all work equipment must be provided with a suitable means of isolating it from its sources of energy. The means of isolation will not be 'suitable' in this context unless it is 'clearly identifiable and readily accessible'. The employer must also ensure that reconnection of any source of energy does not expose anyone to a health and safety risk.

Whilst reg 18 is intended to give effect to para 2.14 of Annex I to Work Equipment Directive, the original text is not qualified by the phrase 'where appropriate'. It is, however, difficult to envisage circumstances where it might not be appropriate, for the purposes of health and safety, to have the means to isolate a power source supplying work equipment.

Regulation 20 provides that an employer 'shall ensure that all work equipment or any part of work equipment is stabilised by clamping or otherwise where necessary for purposes of health or safety'.

Examples, considered above in the context of suitability, include failure to provide retaining brackets for a ladder[88] and failure to provide a suitable run-off board for use when cutting longer materials using a circular saw.[89]

The term 'necessary' is not defined and equivocal guidance was provided in *Robb v Salamis (M&I) Ltd*[90] in this respect. On the one hand, Lord Hope was content to apply the term consistently with 'suitability', ie the duty only being invoked 'when the mischief to be guarded against can be reasonably foreseen'.[91]

Drawing upon wider concerns as to whether PUWER fully implements the Work Equipment Directive, Lord Clyde instead preferred to reserve the question as to whether or not use of the term 'necessary' might not demand a higher standard.[92]

Regulation 21 provides that an employer must ensure that suitable and sufficient lighting is provided at any place where a person uses work equipment. The employer's assessment must take account of the operations to be carried out and overlaps with the duty under reg 8 of the Workplace (Health, Safety and Welfare) Regulation 1992. Consistent with earlier authority under s 5(1) of the Factories Act 1961, the duty is likely to be applied strictly.[93]

---

88 *Robb v Salamis (M&I) Ltd* [2007] 2 All ER 97, HL.
89 *Sherlock v Chester City Council* [2004] EWCA Civ 201, CA.
90 [2007] 2 All ER 97, HL.
91 [2007] 2 All ER 97, HL, per Lord Hope at [23].
92 [2007] 2 All ER 97, HL per Lord Clyde at [46].
93 See *Davies v Massey Ferguson Perkins* [1986] ICR 580, QBD.

The final provisions of general application under PUWER are regs 23 ('Markings') and 24 ('Warnings'). These provisions interrelate and reinforce the other general provisions in a variety of ways.

Regulation 23 strictly requires that all work equipment must be 'marked in a clearly visible manner with any marking appropriate for reasons of health and safety'.

Marking of work equipment may be appropriate for the purposes of reinforcing information, instruction and training. Equally, marking may be appropriate to indicate the presence and purpose of controls or a means of isolation.

Regulation 24(1) further provides that work equipment must incorporate 'any warnings or warning devices which are appropriate for reasons of health and safety'.

Devices may operate to warn operatives about a change of operating conditions or a change in temperature, for example. Mobile work equipment might be fitted with audible and visual warning devices, where used in proximity to other persons.

Regulation 24(1) specifically provides that 'warnings given by warning devices on work equipment shall not be appropriate unless they are unambiguous, easily perceived and easily understood'.

Whilst reg 23 is silent in this respect, markings are unlikely to satisfy the requirements of PUWER unless their meaning or purpose can be readily appreciated by user.

7.13 MOBILE WORK EQUIPMENT

PUWER Part III added new requirements for the employer's management of mobile equipment, which apply in addition to the general provisions of PUWER.

Mobile equipment in this context means,[94]

> 'any work equipment, which carries out work while it is travelling or which travels between different locations where it is used to carry out work...may be self-propelled, towed or remote controlled and may incorporate attachments.'

[94] See ACOP paras 318–323.

In any case where injury results in the following circumstances, specific consideration of the applicable provisions should be made and, particularly in this context, reference made to the ACOP and Guidance:[95]

- an employee being carried upon mobile work equipment – reg 25;
- the rolling over of mobile work equipment – reg 26;
- the overturning of a fork-lift truck – reg 27;
- use of self-propelled work equipment –reg 28;
- use of remote-controlled self-propelled work equipment – reg 29;
- the seizure and/or safeguarding of drive shafts – reg 30.

7.14 POWER PRESSES

PUWER regs32-35 make specific provision for a comprehensive system of examination, inspection, reporting and keeping of information in respect of power presses, their guards and protection devices.

Save as provided at **7.7** above, it is difficult to envisage how Part IV of PUWER might afford greater redress to an injured employee seeking compensation than, for example, regs 5(1) and 11(1).

[95] Regulations 25-30 may operate co-extensively with each other and with the requirements of the Lifting Operations and Lifting Equipment Regulations 1998, SI 1998/2307 where applicable.

CHAPTER 8

WORK EQUIPMENT – SPECIFIC AREAS

8.1 INTRODUCTION

It is not possible to give exhaustive consideration here to the numerous legislative provisions that govern the safe use of specific work equipment or use of work equipment in a specific context.

This chapter concentrates instead upon the Personal Protective Equipment at Work Regulations 1992[1] ('the PPE Regulations'), the Health and Safety (Display Screen Equipment) Regulations 1992[2] ('the Display Screen Regulations') and the Lifting Operations and Lifting Equipment Regulations 1998[3] ('LOLER') on the basis of their wide application.

The above Regulations place particular emphasis on the employer's strict obligation to appropriately assess health and safety risks. In the case of the PPE Regulations, this obligation is only triggered where the risks presented to an employee cannot otherwise be adequately controlled by alternate means – use of PPE is always a course of last resort.

In the case of the Display Screen Regulations, emphasis on the employer's risk assessment obligations reflects the ever-growing dominance of office-based work and extended use of VDU equipment.

Conversely, the additional requirements under LOLER stem more particularly from the immediate, inherent risks commonly presented by use of such equipment.

8.2 THE PPE REGULATIONS

8.2.1 An Overview

The PPE Regulations came into force on 1 January 1993 and are intended to give effect to the PPE Directive 89/656/EEC ('the PPE Directive').

1 SI 1992/2966.
2 SI 1992/2792.
3 SI 1998/2307.

Unlike PUWER, the PPE Regulations are not accompanied by an Approved Code of Practice, though the HSE has issued comprehensive guidance to accompany the same.[4]

A number of consequential amendments to our existing heath and safety legislation were also introduced by the PPE Regulations and, of course, all subsequent legislation must conform to and be construed in any case of ambiguity so as to give effect to the PPE Directive.

Conventional wisdom suggests that strategies built around providing safe places of work, safe equipment and safe systems are likely to prove more effective in promoting health and safety than simple provision of PPE. The former objectives serve to protect everyone within the working environment, whilst the advantages of the latter are limited to the individual. Moreover, an immediate risk of injury is liable to result from any maintenance or system failure should reliance simply be placed upon personal protection.

Given the potential fallibility of the most carefully considered and well-intentioned health and safety scheme, some level of PPE will almost always be necessary to guard against obvious (and not so obvious) risks to vulnerable areas such as the head, face, neck, eyes, ears, respiratory system, skin, arms, hands and feet.

The PPE Regulations impose a comprehensive and co-extensive regime on employers and employees, as summarised below:

Employer

- A duty of assessment.
- A duty to provide suitable PPE.
- A duty to provide compatible PPE.
- A duty to maintain and replace PPE.
- A duty to provide suitable accommodation for PPE.
- A duty to provide information and training.
- A duty to see that PPE is correctly used.

---

4 *Personal Protective Equipment at Work Regulations 1992 – Guidance on Regulations* (2nd edn) L25.

Employee

- A duty to use PPE in accordance with any training and instruction.
- A duty to return PPE to the accommodation provided.
- A duty to report any defect or loss of PPE to the employer.

8.2.2 Interpretation and Application

The term PPE is defined by reg 2(1) as meaning,

> 'all equipment (including clothing affording protection against the weather) which is intended to be worn or held by a person at work and which protects him against one or more risks to his health or safety, and any addition or accessory designed to meet that objective.'

Further to para **7.3** above, the term 'equipment' is interpreted widely. It is neither possible nor desirable to attempt to provide an exhaustive list, but PPE certainly includes the following:

- head protection (helmets, caps, bonnets and hairnets etc);
- hearing protection (ear plugs, mufflers and helmets etc);
- eye and face protection (goggles, masks and face shields etc);
- respiratory protection (filters, respiratory and diving equipment etc);
- hand and arm protection (gloves, wrist protectors and sleeves etc);
- foot and leg protection (shoes, boots, clogs, pads and gaiters etc);
- skin protection (barrier creams and sunscreen etc);
- trunk and abdomen protection (jackets, belts and aprons etc);
- protective clothing (overalls and high-visibility clothing etc);
- personal fall protection (prevention, arrest and braking equipment etc);
- drowning protection (buoyancy aids, jackets and immersion suits etc).

PPE may include waterproof, weatherproof or insulated work clothing in this context where such provision is necessary to protect an employee against adverse climatic conditions that might otherwise adversely affect a person's health or safety.[5]

Further to reg 3(2), the following items are excluded under the PPE Regulations:

- ordinary working clothes and uniforms that do not specifically protect the health and safety of the wearer;
- offensive weapons used as self-defence or as deterrent equipment;
- portable devices for detecting and signalling risks and nuisances;
- personal protective equipment used for protection while travelling on a road (helmets etc);
- equipment used during the playing of competitive sports.

In *Henser-Leather v Securicor Cash Services Ltd*,[6] the Court of Appeal considered the position of a business link employee who was required to collect cash from commercial outlets such as petrol stations. Mr. Henser-Leather was provided with a smoke box, a helmet and instructions that, if confronted, he was to co-operate with any attacker, but his employer did not provide body armour. In the course of his employment, he was challenged by a robber who shot him without warning in the stomach, thereby causing serious injury. The Court of Appeal held that the language of reg 2(1) was sufficiently wide to include body armour.

Further to reg 3(3) as amended, the PPE Regulations are displaced entirely where any of the following instruments, in each case providing a more comprehensive regime, apply:

- Control of Lead at Work Regulations 2002;[7]
- Ionising Radiations Regulations 1999;[8]
- Control of Asbestos Regulations 2006;[9]
- Control of Substances Hazardous to Health Regulations 2002;[10]

---

5 See *Fytche v Wincanton Logistics plc* [2004] 4 All ER 221, HL, per Lord Hoffmann at [15].
6 [2002] EWCA Civ 816, [2002] All ER (D) 259 (May), CA.
7 SI 2002/2676.
8 SI 1999/3232.
9 SI 2006/2739.
10 SI 2002/2677.

- Control of Noise at Work Regulations 2005;[11]
- Construction (Head Protection) Regulations 1989.[12]

The PPE Regulations do not ordinarily apply to merchant shipping activities by virtue of reg 3(1).[13] Whilst the self-employed are subject to the PPE Regulations, otherwise the provisions only apply as between employer and employee. Agency workers, for example, are not covered and there is thus an apparent failure to implement the Art 2(1) and 2(2) of the Temporary Workers Directive.

Regulation 12(1) provides that the Secretary of State for Defence is permitted within some circumstances to exempt the armed forces from the requirements of the PPE Regulations in the interests of national security. Otherwise, the only other partial exemption applies to the conduct of certain policing activities, for example covert surveillance, by virtue of Regulation 4(1A). In this context, the requirement to provide suitable PPE is qualified to the extent that it is 'reasonably practicable' to comply.

8.2.3 Suitability

Regulation 4(1) imposes a strict obligation upon all employers to provide 'suitable' PPE to any employee who may be exposed to a risk to his or her health or safety whilst at work, except where and to the extent that such a risk has been adequately controlled by other means which are equally or more effective.

The term 'risk' is not defined, but the courts ordinarily construe the PPE Regulations consistently with the Manual Handling Operations Regulations 1992 in which the same language is employed. Accordingly, a 'risk' in this context is likely to connote 'a real risk, a foreseeable possibility of injury; certainly nothing approaching a probability'.[14]

In *Henser-Leather v Securicor Cash Services Ltd*,[15] Kennedy LJ observed as follows:

> 'I do not doubt that the risk to someone doing the job which this claimant was doing when he was shot can be to some extent controlled by measures such as parking his van reasonably close to the office from which cash had to be removed, and training employees not to offer resistance and to use the smoke box which was provided. But so long as it remained (as it plainly did) well above the risk to other members of the public going about their daily tasks, it seems to me that the control could not be described as adequate.'

---

11 SI 2005/1643.

12 SI 1989/2209.

13 For which see the Merchant Shipping and Fishing Vessels (Personal Protective Equipment) Regulations 1999, SI/ 1999/2205.

14 *Koonjul v Thameslink Healthcare Services* [2000] PIQR P123 (CA) per Hale LJ at [10].

15 [2002] EWCA Civ 816, [2002] All ER (D) 259 (May), CA.

The Court of Appeal was simply concerned here with the discrete issue of whether the risk of assault by deadly weapon was 'adequately controlled'. A risk cannot be said to be 'adequately controlled' where it remains reasonably foreseeable that injury may result to an employee in the ordinary course of his or her employment.[16] Risk is not judged by reference to level to which the public at large are exposed, whether or not this is broadly comparable. The PPE Regulations can properly be invoked wherever an injured party is able to demonstrate that a reasonably foreseeable risk of injury existed.

The PPE Regulations do not define the term 'suitable' and it is difficult to transpose the guidance offered in relation to reg 4(4) of PUWER, in particular the ruling in *Robb v Salamis (M&I) Ltd*,[17] in this context.

The standard is not absolute, but rather the PPE Directive speaks of the need for PPE to offer protection 'appropriate to the risk involved'. The standard is however clearly a high one.

PPE will not be regarded as 'suitable' if use of alternate PPE would have reduced the prevailing risk further. For example, in *Mitchell v Inverclyde DC*,[18] the injured party succeeded in his claim for personal injury on the basis that non-slip boots were not as effective as studded boots at reducing the risk of falling.

Similarly, in *Toole v Bolton MBC*,[19] the Court of Appeal were content to consider an appeal against the trial judge's finding of contributory negligence on the premise that heavy, Kevlar gloves, which were not impervious to a pinprick injury, would not have been suitable for use as PPE by an employee to clear needle sharps from a waste receptacle. The injured party could not properly attract criticism for falling to utilise PPE if such equipment did not, in any event, comply with the requirements of reg 4(1).

Whatever other considerations may apply, reg 4(3) specifically provides that PPE will not be regarded as 'suitable' unless all of the following conditions are met:

- it is appropriate for the risk or risks involved and the conditions at the place where exposure to the risk may occur;
- it takes account of ergonomic requirements and the state of health of the person or persons who may wear it;

---

16 The standard is certainly not less onerous, for example, than that imposed under the Protection of Eyes Regulations 1974 (revoked by the PPE Regulations) – See *Gerrard v Staffordshire Potteries Ltd* [1995] ICR 502, [1995] PIQR P169, CA.

17 [2007] 2 All ER 97, HL.

18 1998 SLT 1157, 1998 SCLR 191, OH.

19 [2002] EWCA Civ 588, CA.

- it is capable of fitting the wearer correctly, if necessary, after adjustments within the range for which it is designed;
- so far as is practicable, it is effective to prevent or adequately control the risk or risks involved without increasing overall risk;
- it complies with any enactment (whether in an Act or instrument) which implements in Great Britain any provision on design or manufacture with respect to health or safety in any relevant Community directive listed in Sch 1 that is applicable to that item of personal protective equipment.

The latter stipulation thus invites consideration of the Personal Protective Equipment Regulations 2002 ('PPER 2002'), which seeks to approximate or harmonise standards applicable to PPE manufactured or supplied within the EU through implementation of the PPE Directive, as amended.

In this context, it is particularly important to note that Annex II of the PPE Directive, as adopted within Sch 2 of PPER 2002, sets out a number of basic health and safety requirements applicable to all PPE.

On an ergonomic basis,

> 'PPE must be so designed and manufactured that in the foreseeable conditions of use for which it is intended the user can perform the risk-related activity normally whilst enjoying appropriate protection of the highest possible level.'

Accordingly, whilst the PPE regulations do not impose absolute liability in the event of injury during use, the obligation is clearly a strict one, unqualified by issues of cost for example.[20]

PPE must also satisfy, among others, the following basic health and safety requirements:

- It must preclude risks and inherent nuisance factors, such as roughness, sharp edges and projections.
- It must not cause movements endangering the user.
- It must be kept in a clean and hygienic condition and be of a design that enables good hygienic standards, unless it is disposable.
- It must provide comfort and efficiency by facilitating correct positioning on the user and remain in place for the foreseeable period of use.

---

[20] Consistent with the approach under reg 4(1) of PUWER – See *Skinner v Scottish Ambulance Service* [2004] SC 790, [2004] SLT 834, IH.

- It must be as light as possible without undermining design strength and efficiency.
- It must be accompanied by necessary information, eg the name and address of the manufacturer and any technical information.

Unusually, the rigid application of reg 4(1) is perhaps best illustrated by an employment claim, *Lane Group plc v Farmiloe*.[21]

Mr Farmiloe suffered with psoriasis and his condition was exacerbated by the protective footwear, which the risks of his work entailed. There was no other suitable employment for Mr Farmiloe and his employer was unable to obtain alternate, suitable foot protection. In these circumstances, where the employer was liable to be served with an improvement or prohibition notice or to face prosecution under the HSW Act 1974, the EAT held that Mr Farmiloe had not been unlawfully discriminated against, having been dismissed on this basis. HHJ Peter Clark observed as follows:[22]

> '...The significance of *Coxhall* is the holding that there may be cases where an employer is under a duty at law to dismiss the employee so as to protect him from danger. We would go further on the facts of this case, applying *Stark*, and conclude that where an employer cannot comply with the requirements of the PPE Regulations, he will be in breach of his Common Law duty by continuing to employ that individual in breach of the Regulations and in these circumstances, all other avenues having been properly explored will be obliged to dismiss him....'

8.2.4 Compatibility

Regulation 5(1) provides that where the relevant environment, equipment or task presents more than one risk to health or safety, thus making it necessary for an employee to simultaneously use more than one item of PPE, the employer must ensure that such equipment is compatible and continues to be effective against the risk or risks in question. In circumstances where the overall efficiency of a composite PPE system is reduced, when compared with a simple device, an employer will be strictly liable for any resulting injury.

Taking into account ergonomic factors, consistent with approach in *Toole v Bolton MBC* above, an employee will be unlikely to attract criticism for failing to make use of PPE if the system fails complies with reg 5(1).

21 [2004] PIQR P22, EAT.

22 [2004] PIQR P22, EAT, at [43].

8.2.5 Assessment

Regulation 6(1) provides that before choosing any PPE the employer must make an assessment to determine whether such equipment is suitable. A four-stage test is to be applied, the employer being required to:

- identify any risk or risks to health or safety which have not been avoided by other means;
- identify the characteristics which PPE must have in order to be effective against the risks identified above, taking into account any risks which the equipment itself may create;
- compare the characteristics of the PPE available with the characteristics identified above; and
- conduct an assessment as to whether the PPE is compatible with other PPE in use and which an employee may be required to use simultaneously.

Risk in this context means any respect in which it is reasonably foreseeable that the health and safety of a person at work might be affected and will be judged by reference to all matters about which the employer knew or might have ascertained.

Whilst the PPE Regulations are silent in this respect, it is clear from art 4(3) of the PPE Directive that the employer's assessment must have regard to the individual characteristics of the employee, as was the position at common law.

In *Paris v Stepney BC*[23] the plaintiff's claim succeeded because he was known by his employer to have only one sound eye and there was a failure to provide him with appropriate goggles. The court was not satisfied that eye protection ought to have been provided by the employer for anyone working in the same process (based upon the prevailing standards of the day, remembering that this accident occurred in the 1940s) although, obviously, being rendered blind in one eye is a very serious injury for a person with two healthy eyes.

Under reg 4(1), an artificial distinction would no longer be drawn on this basis and a claim would succeed in either case, subject to the injured party establishing the existence of a reasonably foreseeable risk that was not adequately controlled by other means.

In *Pentney v Anglian Water Authority*,[24] the Court held that an employer's failure to provide shatter-proof spectacles to an employee who needed

---

[23] [1951] AC 367, [1951] 1 All ER 42,HL.

[24] [1983] ICR 464, QBD.

glasses constituted a breach of duty at common law. Under reg 6, the employer would now be required to undertake a comprehensive assessment in this and many other everyday instances.

When conducting the second, third and fourth limbs of the assessment process, in particular, it is likely to prove necessary for an employer to seek advice from the relevant supplier or, in more difficult cases, an appropriately qualified specialist.

In all cases, an employer will need to consider the following matters when assessing whether PPE is suitable:

- Is it appropriate for the risks involved and the conditions at the place where exposure to the risk may occur? For example, eye protection designed for providing protection against agricultural pesticides will not offer adequate face protection for someone using an angle grinder to cut steel or stone.

- Does it prevent or adequately control the risks involved without increasing the overall level of risk?

- Can it be adjusted to fit the wearer correctly?

- Has the state of health of those who will be wearing it (including any individual characteristics) been taken into account?

- What are the needs of the job and the demands it places on the wearer? For example, the length of time the PPE needs to be worn, the physical effort required to do the job and the requirements for visibility and communication.

- If more than one item of PPE is being worn, are they compatible? For example, does a particular type of respirator make it difficult to get eye protection to fit properly?

In common with all other health and safety obligations, an employer must undertake a new risk assessment if there is any reason to suspect that the original assessment is no longer valid (for example following an incident, accident or complaint) or where there has been a significant change in circumstances.

8.2.6 Maintenance

Regulation 7(1) provides that all PPE must be maintained (including replaced or cleaned as appropriate) in an efficient state, in efficient working order and in good repair.

Whilst the language employed is identical to reg 5(1) of PUWER, as construed in *Stark v The Post Office*,[25] the House of Lords have since held that the employer's maintenance obligation arising under reg 7(1) relate solely to the specific characteristics of the equipment as PPE, ie whether the equipment retained its integrity and effectiveness against 'identified risks'.

In *Fytche v Wincanton Logistics plc*,[26] the driver of a milk tanker had been supplied with steel toe-capped boots by his employer to reduce the risk of impact injuries. The boots were not supposed to be waterproof and were not intended for use in extreme weather. One of the boots had a tiny hole in it when supplied. Mr Fytche had got out of his cab to dig the tanker out of snow and he suffered mild frostbite to his toe due to water penetration. The employer's instructions to deal with such a situation were for a driver to use the telephone in the cab and await assistance.

The House of Lords held, by a majority of three to two, that the presence of a small hole did not render the boot unsuitable for the purpose for which it had been supplied and did not constitute a failure to maintain the same.

The seemingly irrational and illogical nature of the reasoning of the majority is eloquently encapsulated by the dissenting judgment of Baroness Hale:[27]

> '…The issue in this case…is who should bear the risk that the boots supplied for a particular reason turn out to have an incidental defect which causes the employee injury while he is at work. I have no difficulty with the conclusion that the employer rather than the employee should bear that risk. There are good policy reasons for imposing strict liability on employers for many of the injuries which their employees suffer at work. The overall object of the legislation is to protect the health and safety of workers: if this fails and they suffer injury, strict liability means that they are compensated for that injury without the need for slow and costly litigation such as this. I appreciate that we have not yet reached the point where there is strict liability for every injury suffered by a worker in the course of his employment, but I see no need to bring in limitations which are not in the statutory language and could, as illustrated above, lead to some very surprising conclusions. I venture to suggest that a non-lawyer would find it odd indeed that Mr Fytche would have recovered damages if his employer had also thought the boots should protect against a weather risk but does not do so because his employer had a different risk in mind ….'

Quite apart from compelling policy reasons, it is worth emphasising that use of PPE is mandatory (see reg 10(2) below), selection and supply is the exclusive provenance of the employer (Mr Fytche was not free to turn up

---

25 [2000] PIQR P105.

26 [2004] 4 All ER 221, HL.

27 [2004] 4 All ER 221, HL, at [70].

in his own boots) and failure to wear the defective equipment supplied to him would have given his employer grounds for dismissal.

Whilst *Fytche v Wincanton Logistics plc* presently represents an accurate statement of law relating to maintenance of PPE, it remains to be seen whether the House of Lords will be persuaded to revisit this issue or whether a collateral attack upon the ruling might be made by placing reliance upon the Employers' Liability (Defective Equipment) Act 1969. Subject to acceptance that boots constitute 'equipment' for these purposes, there is no reason in principle why the two instruments should not apply co-extensively.

Advisers will commonly need to examine the employer's practices at a fairly general level to ascertain where and how the relevant system broke down. In particular, it will often be helpful to consider whether:

- PPE was kept clean and in good repair, following the manufacturer's maintenance schedule (including recommended replacement periods and shelf lives) as appropriate?

- PPE was readily available and replaced as appropriate?

The employer's duty of maintenance is coupled with a duty, arising under reg 8, to provide appropriate accommodation for PPE when it is not being used. Advisers should examine whether PPE was "looked after" in the sense of being properly stored when it was not being used, for example in a dry, clean cupboard, or in the case of smaller items, such as eye protection, in a suitable receptacle.

Whilst the obligation arising under reg 8 is strict, it is highly unlikely that direct reliance would be placed upon such a breach in the context of a personal injury action. The duty is however relevant, on an evidential basis, in a number of contexts, for example, as to:

- the employer's general observance of statutory requirements;

- the circumstances in which PPE may have been damaged (eg where the relevant equipment is no longer available for inspection or where operative error is alleged);

- the circumstances in which PPE may have been lost;

- the relative blameworthiness of the employee where use of PPE is not observed (eg item left at home) and/or where the employee has failed to report loss or damage.

8.2.7 Information, Instruction and Training

Regulation 9((1) provides that the employer must ensure that an employee using PPE receives adequate and appropriate information, instruction and training as to the following:

- the risk or risks which the PPE will avoid or limit;
- the purpose for which and the manner in which PPE is to be used;
- any action to be taken by the employee to ensure that the personal protective equipment remains in an efficient state, in efficient working order and in good repair as required by reg 7(1) above.

Such information, instruction and training is specifically directed to issues of health and safety and the adequacy or appropriateness of the employer's steps fall to be considered accordingly. Relevant factors, forming part of the employer's assessment, are likely to include:

- nature of PPE (simple, complex, multi-risk or combined etc);
- nature of task (including complexity, conditions and duration etc);
- risk factors (including likelihood and likely severity of harm);
- age, sex and/or experience (young and pregnant workers etc);
- individual characteristics (disability or language considerations etc).

In like manner to reg 8 of PUWER, it is a mandatory requirement that information and instruction in connection with use of PPE is 'comprehensible to the persons to whom it is provided'. Similarly, the employer's duty to provide information, instruction and training in this context is an strict and continuing one. It is no answer for an employer to suggest that training or instruction was too costly or impracticable.[28]

There is a highly significant respect in which the PPE Regulations go further than PUWER on the issue of information, instruction and training. Regulation 9(3) provides that an employer must where appropriate and at suitable intervals organise demonstrations relating to the wearing of PPE.

This provision is unique in its express recognition that training involving active demonstration is always preferable to other means, the promotion

[28] See *Pennington v Surrey County Council* [2006] EWCA Civ 1493, [2007] PIQR P11 in the context of PUWER.

of health and safety awareness amongst employees is a continuing duty and that good practice should be encouraged through repetition over time.

8.2.8 Enforcement

It is against this background that reg 10(1) of the PPE Regulations must be carefully considered. It provides that:

> 'Every employer shall take all reasonable steps to ensure that any personal protective equipment provided to his employees by virtue of regulation 4(1) is properly used.'

The obligation is strict, continuing and entirely discrete from the duty to provide adequate and appropriate information, instruction and training. The employer will, of course, be expected to take into account all of the factors that are relevant in the context of implementing regs 6 and 9 above when determining what steps are reasonably required in this context.

The burden upon the employer is likely to be a difficult one to discharge in all cases. In *Henser-Leather v Securicor Cash Services Ltd*,[29] the Court of Appeal roundly rejected any comparison between the duties arising under reg 10 and the position at common law.

In *Nolan v Dental Manufacturing Co*[30] for example, the defendant was under a statutory duty to provide goggles, but no such provision was made. While Mr Nolan was sharpening a hand tool, a chip flew off causing him to suffer an eye injury. Mr Nolan failed in his action for damages for breach of statutory duty because although a breach was established, the court was not satisfied that he would have worn goggles had they been provided.

In *Henser-Leather v Securicor Cash Services Ltd*, by contrast, Kennedy LJ observed as follows:[31]

> '...Regulation 10 placed obligations on employer and employee in relation to the wearing of such equipment...The defendants took no steps to comply with Regulation 10 (1) so it is, in my judgment, impossible to contend that the claimant would not have complied with his obligations under Regulation 10 (2) had the body armour been supplied and had those steps been taken. The judge found that if body armour had been supplied and it had been left to him to decide whether or not to wear it, he would not have done so. That is a totally different situation ....'

Whilst reg 10(2) also places the employee under a duty to use PPE in accordance with the training and instruction provided by the employer,

---

[29] [2002] EWCA Civ 816, [2002] All ER (D) 259 (May), CA.

[30] [1958] 2 All ER 449.

[31] [2002] EWCA Civ 816, [2002] All ER (D) 259 (May), CA, at [23]-[24].

there will rarely be circumstances in which, at the very least, primary liability is not established unless all of the employer's obligations have been fully complied with. In particular, the employer should be required to first prove each of following matters on the balance of probability:

- the PPE was suitable for the purpose it was provided;
- the employee was provided with adequate and appropriate information, instruction and training concerning use of the PPE;
- such information, instruction and training was reinforced as appropriate;
- loss or damage (in the event that the employee failed to report the same) was not caused or contributed to by a failure to provide suitable accommodation for the PPE; and
- all reasonable steps to ensure use of PPE by the relevant employee were taken.

All PPE should be 'CE' marked so as to demonstrate compliance with the requirements of the PPER 2002. The CE marking signifies that the PPE satisfies certain basic safety requirements and in some cases will have been tested and certified by an independent body. Advisers should note that provision by an employer of PPE that does not comply with the requirements of PPER 2002 represent an automatic breach of reg 4(1) by virtue of reg 4(3)(e).

8.3 THE DISPLAY SCREEN REGULATIONS

8.3.1 The Display Screen Regulations – An Overview

The **Health and Safety (Display Screen Equipment) Regulations 1992**[32] ('Display Screen Regulations') came into force on 1 January 1993 and are intended to give effect to Council Directive 90/270/EEC on the minimum safety and health requirements for work with display screen equipment ('the Display Screen Directive').

Of the European influenced health and safety provisions, the Display Screen Regulations are one of few instruments specifically targeted at a particular type of equipment and activity (contrast the other six-pack Regulations which are of generic application across a wide range of roles or tasks).

The Regulations are not overly prescriptive in the sense of containing detailed technical specifications or lists of approved equipment, but rather

[32] SI 1992/2792.

set more general objectives. In broad summary, employers are expected to consider the following matters where extended use of VDU equipment is a feature within the relevant undertaking:

- the whole workstation including equipment, furniture and the work environment generally;
- the job being done; and
- any special needs of individual staff.

Whilst it is not an express requirement, the Regulations are clearly premised on the assumption that the employer will consult effectively with the individual employee (and safety representatives where appropriate) as part of this assessment process. Where risks are identified, the employer is thereby placed under a strict duty to:

- analyse workstations, and assess and reduce such risks;
- ensure workstations meet minimum requirements;
- plan work so there are breaks or changes of activity;
- on request arrange eye tests, and provide spectacles if special ones are needed;
- provide health and safety training and information.

The health risks presented by extended use of VDU are well documented. In most cases, advisors will find themselves working back from the diagnosis of a relevant condition. Once it is established that a client's injury is probably work-related (based upon medical opinion) it is then very difficult for an employer to argue that the employee's problems were not thus caused or materially contributed to by any established breach of the Regulations.[33]

Whilst the Regulations are not accompanied by an approved code of practice, the HSE has issued comprehensive guidance to accompany the same.[34] In this context, perhaps as much as in any other, advisers (and the courts) should carefully considered the HSE Guidance when evaluating the extent to which an employer can be said to have complied fully with the requirements of the Regulations.

---

[33] *Fifield v Denton Hall* [2006] Lloyd's Rep (Med) 251 per Wall LJ at [48].

[34] *Work with Display Screen Equipment – Guidance on Regulations* (2nd edn) L26.

Case law in this context is still fairly limited and, in most instances, the court's focus is principally limited to issues of medical causation.[35] A number of general considerations, however, emerge:

- In this context, as least, the courts show a general disinclination to formulaic 'tick-box' exercises relating to assessment, instruction and training. These tend to defeat the learning or behavioural objectives that are promoted by effective employee participation.

- The actual day-to-day practices within the workplace need to be closely scrutinised in each case. Where an employee has effectively been left to his or her own devices, the employer can expect robust censure notwithstanding any paper evidence of appropriate assessment and risk management.

- The Regulations must to be considered on a holistic basis, i e they are intended to interrelate and entail an assessment of all known risks within the workplace.

- Where relevant breaches are established (in particular as to working practices), it will be almost impossible for an employer to successfully to mount a technical defence based upon legal causation, i e by alleging that the employee was 'the author of his or her own misfortune' or that, if appropriately implemented, health and safety training and information would have been disregarded come what may.

8.3.2 Interpretation and Application

Regulation 1(2) of the Health and Safety (Display Screen Equipment) Regulations 1992 provides a number of fairly straightforward definitions. The terms 'operator' and 'user' are defined by reference to 'habitual use' of display screen equipment as a 'significant part' of his or her normal work.

Whether an employee's use of display screen equipment is 'habitual' and a 'significant part' of his or her normal work will be a question of fact in each case. Bearing in mind the general purpose of the Regulations, the duration of display screen activity might logically be regarded as a 'significant part' in any case where the medical or ergonomic evidence suggests that it give rise to a reasonably foreseeable health risk in the circumstances.

Regulation 8(1) provides that the Secretary of State for Defence may exempt the armed forces from the requirements of the Regulations in the interests of national security. Otherwise, the Regulations apply

[35] For example, see *Fifield* above, *MacPherson v Camden LBC* (unreported) 16 September 1999, QBD and *Gallagher v Bond Pearce* [2001] 6 QR 15.

co-extensively to use of all display screen equipment by employees, the self-employed and temporary workers, save as provided below:

- drivers' cabs or control cabs for vehicles or machinery;
- display screen equipment on board a means of transport;
- display screen equipment mainly intended for public operation;
- portable systems not in prolonged use;
- calculators, cash registers or any equipment having a small data or measurement display required for direct use of the equipment;
- window typewriters.

8.3.3 Analysis of Workstations

Regulation 2 imposes a comprehensive, multi-factorial risk assessment and risk reduction regime. Employers are required to:

- make a suitable and sufficient analysis of those workstations which:
  - (i) (regardless of who has provided them) are used for the purposes of his undertaking by users, or
  - (ii) have been provided by him and are used for the purposes of his undertaking by operators;
- assess the health and safety risks to which those operators or users are exposed in consequence of that use;
- reduce those risks to the lowest extent reasonably practicable, and to review (and where necessary change) any assessment as appropriate.

Advisers should note that Art 2(3) of the Display Screen Directive, to which reg 2 is intended to give effect, is not qualified by the words 'to the lowest extent reasonably practicable', but instead refers to 'appropriate measures'. The Display Screen Directive thus arguably imposes a higher standard, directly applicable in respect of any state employer.

8.3.4 Requirements for Workstations

Regulation 3 lays down minimum requirements for workstations. Employers must ensure that workstations comply with the requirements set out in Schedule to the Regulations (as laid out in the Annex to the Display Screen Directive) under the following headings:

- equipment (including the display screen, keyboard, work desk or work surface and work chair);

- environment (including space, lighting, reflections and glare, noise, heat, radiation and humidity); and
- interface between computer and operator/use (including software and task considerations).

8.3.5 Daily Work Routines

Regulation 4 makes specific provision for varying employee exposure. The employer is strictly obliged to plan the activities of display screen users so that that daily work on such equipment is periodically interrupted by breaks or changes of activity so as to reduce the workload and thus limit the associated risk of injury.

It is not sufficient for an employer simply to provide written advice or training in this context. The court will be highly critical where employees are simply left to devise or implement safe working practices for themselves.[36]

8.3.6 Eyes and Eyesight

Regulation 5 provides for an elective eye care regime (in the sense that the employee cannot be compelled to participate). The employer must ensure that employees using display screen equipment are provided:

- with initial eye and eyesight tests on request;
- with subsequent eye and eyesight tests at regular intervals;
- with additional eye and eyesight tests on request, where the employee concerned is experiencing visual difficulties which might reasonably be considered to be caused by work on display screen equipment; and
- with appropriate special corrective appliances, where normal corrective appliances cannot be used and any eye and eyesight tests carried out on the users concerned in accordance with regulation 5 show such provision to be necessary.

8.3.7 Training and Information

Regulations 6 and 7 impose a strict and continuing obligation on employers to devise, implement and maintain a comprehensive regime for the provision of adequate health and safety training and information for all users of display screen equipment. Regulation 6(1A) makes it clear

---

36 *Fifield v Denton Hall* [2006] Lloyd's Rep (Med) 251.

that the employer's duty operates prospectively, ie before the employee commences such work and thus before exposure to the accompanying risks.

8.4 LIFTING EQUIPMENT (LOLER)

8.4.1 LOLER – An Overview

The Lifting Operations and Lifting Equipment Regulations 1998[37] ('LOLER') came into force on 5 December 1998. The Regulations revoke and replace a large number of earlier provisions and are intended to give effect to para 3.2 of Annex I ('minimum requirements for work equipment for lifting loads') and para 3 of Annex II ('provisions concerning the use of work equipment for lifting loads') of the Work Equipment Directive.

The requirements are to be read in conjunction with PUWER, particularly Part III concerning use of mobile work equipment. Also, regard should always be had to the approved code of practice and guidance accompanying LOLER.[38]

In general terms, LOLER prescribes that 'lifting equipment' provided for use at work must be:

- of adequate strength and stability;
- marked to indicate safe working loads and other aspects of safe use;
- positioned and installed to minimise any risks;
- used safely, ie the relevant task is planned, organised and performed by competent people; and
- subject to ongoing and thorough examination and, where appropriate, inspection by competent people (with appropriate records kept).

8.4.2 Interpretation and Application

The term 'lifting equipment' is defined by reg 2(1) as meaning 'work equipment for lifting or lowering loads and includes its attachments used for anchoring, fixing or supporting it', whilst the term 'lifting operation' is defined simply by reg 8(2) as 'an operation concerned with the lifting or lowering of a load'.

---

37 SI 1998/2307.

38 Safe Use of Lifting Equipment – Lifting Operations and Lifting Equipment Regulations 1998 (L113) approved and issued pursuant to s 16(1) of the HSW etc 1974.

Further to para **7.3** above, the term 'equipment' is interpreted widely. The Regulations cover a range of items and devices including, cranes, fork-lift trucks, lifts, hoists, mobile elevating work platforms and vehicle inspection platform hoists. The definition also includes lifting accessories or components such as ropes, chains, slings, shackles and eyebolts.

Escalators are not covered by LOLER, but are instead the subject of specific consideration under reg 19 of the Workplace (Health, Safety and Welfare) Regulations 1992.[39] Neither are conveyor belts nor winch devices, for example, which operate on a horizontal level.

The Regulations apply in whichever industry such equipment is utilised and the responsibilities apply to each of the following user groups (in some cases co-extensively, as with PUWER above):

- an employer in respect of lifting equipment provided for use or used by his employee – reg 3(2);
- a self-employed person, in respect of lifting equipment used by him at work – reg 3(3)(a);
- anyone exercising control of (i) lifting equipment, (ii) anyone using, supervising or managing the use of lifting equipment or (iii) the way in which lifting equipment is used at work – reg 3(3)(b).

The Regulations also cover places wherever the HSW etc Act 1974 applies, thus including factories, offshore installations, agricultural premises, offices, retail outlets, hospitals, hotels and leisure facilities.

In like manner to the PPE Regulations, LOLER does not apply to merchant shipping activities by virtue of reg 3(6)-(10).[40] Similarly, reg 12(1) provides that the Secretary of State for Defence may exempt the armed forces from the requirements of the PPE Regulations in the interests of national security.

8.4.3 Strength and Stability

One of five key requirements in the context of any personal injury claim arising from the use lifting equipment, reg 4 imposes a strict obligation on employers to ensure that:

- lifting equipment is of adequate strength and stability for each load, having regard in particular to the stress induced at its mounting or fixing point; and

---

[39] SI 1992/3004.

[40] For which see the Merchant Shipping and Fishing Vessels (Lifting Operations and Lifting Equipment) Regulations 2006, SI 2006/2184.

- every part of a load and anything attached to it and used in lifting it is of adequate strength

Consistent with the approach under PUWER to 'suitability', the 'adequacy' of lifting equipment in this context is likely to be determined with reference to the intended or any reasonably foreseeable use.

Naturally, any acute failure of lifting equipment would in all likelihood give rise to a discreet and irresistible claim under reg 5(1) of PUWER following *Stark v The Post Office*.[41]

Otherwise, reg 4 reinforces the employer's obligations arising under regs 4(1)-(3) and 20 of PUWER.

Misuse or use of inappropriate lifting equipment is likely to also constitute a breach of reg 8(1) below.

8.4.4 Lifting Persons

Whenever lifting equipment is used for lifting persons, thereby involving an acute health and safety risk, the requirements of reg 5 of LOLER apply in addition and in priority to all other obligations. It provides that every employer must ensure that lifting equipment for lifting persons:

- is such (presumably meaning selected, constructed or adapted etc in this context) as to prevent a person using it being crushed, trapped or struck or falling from the carrier;

- is such as to prevent a person using it, while carrying out activities from the carrier, being crushed, trapped or struck or falling from the carrier;

- has suitable devices to prevent the risk of a carrier falling, and, if the risk cannot be prevented for reasons inherent in the site and height differences, the employer must ensure that the carrier has an enhanced safety coefficient suspension rope or chain which is inspected by a competent person every working day; and

- is such that a person trapped in any carrier is not thereby exposed to danger and can be freed.

Where activities are being carried out from the carrier (i e where the lifting equipment is not simply being used to lift or lower the employee, but also in effect as a place of work), the employer's strict duty is qualified to the extent that it is 'reasonably practicable' to prevent such occurrences.

---

41 [2000] PIQR P105, CA.

Advisers should note that para 3.2.4 of Annex I to the Work Equipment Directive, to which reg 5 is seemingly intended to give effect, does not provide any such qualification.

Paragraph 128 of the ACOP specifically provides that the raising and lowering of people should only be undertaken using work equipment that is specifically designed for such purposes (eg cherry pickers), save only in exceptional circumstances, ie when it is not practicable to gain access by alternative means, and then only when all necessary precautions have been taken.

8.4.5 Positioning and Installation

Regulation 6(1) provides that every employer shall ensure that lifting equipment is positioned or installed in such a way as to reduce to as low as is reasonably practicable the risk:

- of the lifting equipment or a load striking a person; or
- from a load drifting, falling freely or being released unintentionally.

Regulation 6(1) also provides that lifting equipment must otherwise be positioned or installed so as to ensure that it is 'safe'. This duty is strict and unqualified. If the positioning or installation of lifting equipment causes or contributes to an alternate risk, liability will attach for any resulting injury.

Advisors should note that the expression 'safe' has been construed strictly, in the context of s 29 of the Factories Act 1969, without reference to reasonable foreseeability.[42]

Regulation 6(2) additionally provides that 'every employer shall ensure that there are suitable devices to prevent a person from falling down a shaft or hoistway'.

8.4.6 Markings

Further to the general requirement under reg 23 of PUWER to provide appropriate health and safety markings, reg 7 of LOLER specifically provides that employers must ensure that:

- machinery and accessories for lifting loads are clearly marked to indicate their safe working loads;
- where the safe working load of machinery for lifting loads depends on its configuration, either the machinery is clearly marked to

---

42 See *Larner v British Steel plc* [1993] 4 ALL ER 102, [1993] IRLR 278, CA.

indicate its safe working load for each configuration, or information which clearly indicates its safe working load for each configuration is kept with the machinery;

- accessories for lifting are also marked in such a way that it is possible to identify the characteristics necessary for their safe use;
- lifting equipment which is designed for lifting persons is appropriately and clearly marked to this effect; and
- lifting equipment which is not designed for lifting persons but which might be mistakenly so used is appropriately and clearly marked to the effect that it is not designed for lifting persons

Whilst reg 7 is silent in this respect (contrast PUWER and the PPE Regulations), it is clear that markings will not be appropriate in this context unless readily comprehensible to all those concerned by the use of lifting equipment (eg take into account individual characteristics as appropriate).

8.4.7 Organisation

Regulation 8(1) is likely to be the principle duty of assistance in most claims arising out of an alleged breach of LOLER. It is an extensive and strictly applicable provision. It entails that every lifting operation involving lifting equipment, without exception or qualification, is:

- properly planned by a competent person;
- appropriately supervised; and
- carried out in a safe manner.

In *Delaney v McGregor Construction (Highlands) Ltd*,[43] the defendant had used a fork lift truck to unload steel roads from a lorry in the absence of a crane or sling. The rods fell severely injuring Mr Delaney whilst he assisted with the operation. Lady Paton observed as follows:[44]

> '...In my view, the pursuer has established a breach of regulation 8....in November 1999 the safe way of unloading a bundle of long flexible steel rods weighing about a tonne was by crane and slings or some similar arrangement. The method involving the JCB with forks was in my view improvised, unsatisfactory, not easily controllable, and accordingly risky for those in the vicinity. I am therefore unable to accept that the operation was properly planned by a competent person, or that it was carried out in a safe manner...

---

43 [2003] Rep LR 56, OH.

44 [2003] Rep LR 56 at [91]-[92].

> ...The defenders failed to satisfy me that it was not reasonably practicable to provide a crane and slings, or a similar arrangement. A breach of regulation 6(1)(b)(ii) has also in my view been established...'

The defendant failed to explain or adduce evidence as to why it had used a fork lift truck in such circumstances. Consistent with an extensive body of judicial authority, the court proceeded on the basis that the burden of demonstrating compliance with the LOLER rested with the defendant.

8.4.8 Examination, Inspection and Reporting

Regulation 9 provides for a comprehensive regime of examination and testing in respect of lifting equipment, whilst reg 10 makes specific provision for reporting of defects and consequential restrictions on the use of equipment prior to rectification. The information obtained in furtherance of regs 9 and 10 must be retained in accordance with reg 11.

Further to the observations in respect of PUWER, a breach of the above Regulations is only likely to be of evidential significance in the context of a personal injury action, i e evidence of the employer's general compliance and record keeping (or a lack of it) may serve bolster or undermine a claim as the case may be.

CHAPTER 9

MANUAL HANDLING

9.1 BEFORE 'MANUAL HANDLING'

Prior to January 1993, the term 'manual handling' was little recognised. Lawyers acted for or against injured people who had suffered what were known as 'lifting injuries'. Lifting as a term covered lifting, carrying or moving loads. In practice the weight of the load that the injured person had been dealing with at the time of their accident was the only thing that usually dictated the outcome of any case.

The law, as far as it went, was decided on judicial interpretation of cases founded on a wide range of statutes and regulations and common law negligence. The most common basis of a claim was an alleged breach of s 72 of the Factories Act 1961. However for claims not concerning accidents in factories, s 23 of the Offices, Shops and Railway Premises Act 1963 was often relied upon. If that did not apply there were other Acts and regulations you might consider, including the Agriculture (Lifting of Heavy Weights) Regulations 1959[1]or the Construction (General Provisions) Regulations 1961[2] that might be relied upon. There was simply no standard for workers across the board and many had little or no statutory protection at all.

9.1.1 Early Statutory Protection

The most important part of the old law was found in s 72 of the Factories Act 1961 which stated: 'A person shall not be employed to lift, carry or move any load so heavy as to be likely to cause injury to him'. It was that simple.

Three Court of Appeal cases show clearly how those words were interpreted. In *Kinsella v Harris Lebus Ltd*,[3] Kinsella was injured while attempting to lift a weight of 145lb. at his employer's factory. The Court

---

1 SI 1959/2793.

2 SI 1961/1580.

3 (1964) 108 Sol Jo 14.

of Appeal held that the Factories Act was designed to give protection against excessive weight only and 145lb. was not likely to cause injury to a man of experience.

In *Peat v J Muschamp & Co*[4] the Court of Appeal held that where an employee, knowing that help is readily available, lifts a heavy load and suffers injury, his employer is not liable under the Factories Act. The court also held that Peat (a man who only had only one leg) in lifting a load of 65lbs was not employed to carry or move a load that was so heavy as to be likely to cause injury to him within s 72, since he 'had been told, whenever he wanted help, to get it'.[5]

Perhaps the most surprising decision of all is Bailey v Rolls Royce (1971) Ltd.[6] Bailey was employed as a paint-sprayer by Rolls Royce. Bailey was occasionally required to lift heavy objects in the course of his employment. In November 1976 he injured his back in lifting a heavy object. Bailey had had previous back trouble although not serious. The employer's records showed that Bailey had strained his back twice in April 1976 and was off work in May and June 1976 with backache. Bailey's doctor gave evidence that he was a chronic back sufferer and he would have warned Bailey's employers about it if asked to do so.

The Court of Appeal held that the phrase 'likely to cause injury to him' (the employee) in s 72 meant that injury must be more probable than not and did not impart any requirement of foreseeability on the part of the employer. They said this was a borderline case but found on the evidence that the employer was neither negligent nor in breach of s 72. He was moving a weight of 176lb.

Clearly there were serious problems with the law, which was weighted heavily against employees. Fortunately major change arrived in the form of a European directive, and now such harsh decisions are unthinkable. The European directive has completely revolutionised the way the courts are obliged to approach 'lifting injuries' starting with the old established focus on the weight of the object lifted.

9.2 MANUAL HANDLING DIRECTIVE

The Manual Handling Directive (90/269/EEC) was the fourth individual directive made on 29 May 1990 under Art 16 (1) of the Framework Directive (89/391/EEC).[7] It provides for minimum health and safety

---

4 (1969) 7 KIR 469.

5 Per Davies LJ.

6 [1984] ICR 688.

7 The Manual Handling Directive is implemented by the Manual Handling Operations Regulations 1992, SI 1992/2793 in England and Wales, and the Manual Handling Operations Regulations (Northern Ireland) 1992, SR 1992/535 in Northern Ireland. See further below.

requirements for the manual handling of loads where there is a risk particularly of back injury to workers. However, as the term 'particularly' suggests it covers much more than just back injuries.

The definition in Art 2 states that:

> 'manual handling of loads means any transporting or supporting of a load, by one or more workers, including lifting, putting down, pushing, pulling, carrying or moving of a load, which, by reason of its characteristics or of unfavourable ergonomic conditions, involves a risk particularly of back injury to workers.'

In other words if there is physical effort involved the Directive applies. Key too is the word 'ergonomic' which dismisses the possibility that weight alone can be the only factor in determining liability.

Article 3 requires that the employer shall take:

> 'appropriate organisational measures, or shall use the appropriate means, in particular mechanical equipment, in order to avoid the need for the manual handling of loads by workers.'

In other words do not move the load but rather mechanise its movement.

It goes on to say that where the need for the manual handling cannot be avoided, the employer shall take the appropriate organisational measures, use the appropriate means or provide workers with such means in order to reduce the risk involved in the manual handling of such loads, having regard to a long list of matters set out in Annex I of the Directive.

Article 4 looks at the organisation of workstations. It also refers to situations where the need for manual handling cannot be avoided reminding us of the fact that 'avoidance' is the starting point. There is a requirement to organise workstations in such a way as to make handling as safe and healthy as possible.

It also requires the employer to assess, in advance if possible, the health and safety conditions of the type of work involved, and in particular examine the characteristics of loads, again taking account of Annex I. Following assessment the duty is to take care to avoid or reduce the risk by taking appropriate measures, considering in particular the characteristics of the working environment and the requirements of the activity, again using Annex I as a guide.

Annex II deals with individual risk factors that must be considered with Art 6 of the Directive. It recognises that we are all different. It points out that a worker may be at risk if he or she:

- is physically unsuited to carry out the task in question,

- is wearing unsuitable clothing, footwear or other personal effects,
- does not have adequate or appropriate knowledge or training.

Article 6 covers the provision of information for, and training of, workers and the need to keep them or their representatives informed of all measures to be implemented, with regard to the protection of safety and of health under the Directive. It also covers the duty on employers to ensure that they receive general indications and, where possible, precise information on the weight of a load and the centre of gravity of the heaviest side when a package is eccentrically loaded. It also requires that workers additionally receive proper training and information on how to handle loads correctly and the risks they might be open to particularly if these tasks are not performed correctly, having regard to Annexes I and II.

Finally Art 7 requires consultation of workers and workers' participation to comply with Art 11 of the Framework Directive on matters covered by the Manual Handling Directive, including the Annexes.

9.2.1 Direct Effect

As with all directives there is 'direct effect' against an emanation of the state. That has applied to manual handling since 29 May 1990. In *Peck v Chief Constable of Avon and Somerset*[8] Recorder Harrop held, giving judgment for Peck, that Directive 90/269 applied to the defendant as an 'emanation of the State'. There was a breach of the Directive in that Peck, a police officer, was allowed to lift a protestor on his own, or more specifically he had not been instructed to lift only with a fellow officer, and had not received any training about lifting in such circumstances.

In other instances the start date for the directive to come into play was 1 January 1993. That is the date the current UK Regulations took effect.

9.2.2 Purposive Approach

The Manual Handling Directive remains the key to ensuring that the Manual Handling Operations Regulations 1992 are applied purposively. In *Cullen v North Lanarkshire Council*[9] the Inner House held that purposive interpretation of the regulations is always required and any operation *involving* manual handling is covered by the Directive and Regulations. The Regulations applied to a job when a workman was throwing fencing off the back of a lorry and lost his footing and fell. The accident was not caused by the weight of the load but by poor footing. The fact that any operation *involving* manual handling is covered is clearly

---

[8] [2000] 5 CL 330.
[9] [1998] SC 451.

of considerable importance as it vastly increases the number of accidents where the Regulations are applicable.

However the Directive is not a charter for defeating the Regulations. The Regulations must provide protection as the Directive intended but they need do no more. In *Sussex Ambulance NHS Trust v King*[10] the claimant was injured in the course of his employment when he and a colleague answered a call to collect a patient from his home and take him to hospital.

The patient was an elderly man who was upstairs in bed in a cottage that had a narrow and steep stairway with a bend in it. As they carried the patient down the stairs King was injured. The Court of Appeal accepted that the task was clearly hazardous but that there was no alternative means of removing the patient short of involving the fire brigade to lift him out through a removed window with a crane. At first instance the judge found that the trust had breached Art 3.2 of the Manual Handling Directive by failing to train its employees to give serious consideration to the alternative of using the fire brigade and by treating it very much as an option of last resort.

The Court of Appeal held that there was liability under the Directive or under the Regulations. Calling the fire brigade would have been an *appropriate measure* in this case. That term was used in the Directive. It was argued that a requirement to take *appropriate measures* imposed a higher duty than the requirement under the Regulations based on a defence of reasonable practicability. In reality there was no difference and the appeal was successful.

9.3 MANUAL HANDLING REGULATIONS

The Manual Handling Directive is implemented by the Manual Handling Operations Regulations 1992[11] in England and Wales, and the Manual Handling Operations Regulations (Northern Ireland) 1992[12] in Northern Ireland. They both came into force on 1 January 1993 with immediate effect.

To know if these Regulations are relevant at all an employer must have identified manual handling operations in his undertaking. That should have happened when the general risk assessment required by reg 3 of the Management of Health and Safety at Work Regulations 1999 took place.

If the general assessment indicates the possibility of risks from manual handling of loads then they have also identified the need to apply the

---

10 [2002] EWCA Civ 953.

11 SI 1992/2793.

12 SR 1992/535.

requirements of the Manual Handling Operations Regulations 1992. The Management of Health and Safety at Work Regulations usually do no more than that as far as manual handling is concerned.

9.3.1 What is Excluded?

Regulation 2 (1) of Manual Handling Operations Regulations 1992 limits the definition of 'injury' to exclude injury caused by any toxic or corrosive substance which:

(i) has leaked or spilled from a load;

(ii) is present on the surface of a load but has not leaked or spilled from it; or

(iii) is a constituent part of a load.

9.3.2 What is Covered?

Regulation 2 (1) of Manual Handling Operations Regulations 1992 confirms that 'load' includes any person and any animal. It defines 'manual handling operations' very clearly as:

> 'any transporting or supporting of a load (including the lifting, putting down, pushing, pulling, carrying or moving thereof) by hand or by bodily force.'

It imposes the duties under the Regulations on an employer in respect of his employees and also on a self-employed person in respect of himself.

Regulation 3 limits the application of the Regulations by saying that they:

> 'do not apply to or in relation to the master or crew of a sea-going ship or to the employer of such persons in respect of the normal ship-board activities of a ship's crew under the direction of the master.'

Everything else is covered. All employees are covered, even those who are temporary or 'ad hoc' as was held by His Honour Judge Hawkesworth QC in *Whitcombe v Baker*.[13]

---

[13] (unreported) 9 February 2000, QBD.

An injury can be a single event accident or a cumulative injury over time. See for example *Fiona Stone v Metropolitan Police Comr*[14] and *Knott v Newham Healthcare NHS Trust*[15] in neither of which was there any specific accident.

Causation is the major area at issue in such cases. It is also important to show that it was a manual handling operation that caused the injury. This difficult was illustrated very clearly in the case of *Hughes v Grampian Country Food Group Ltd.*[16]

In Hughes the Inner House of the Court of Session was concerned with a process worker whose job included trussing wings and legs of chicken carcasses using elastic strings. She developed Carpal Tunnel Syndrome in her left wrist that was aggravated by her work. She claimed damages for the aggravation alleging a breach of reg 4 of the Manual Handling Operations Regulations 1992.

The job involved picking up a carcass and placing it on the workbench in front of her. While the carcass was on the workbench she manipulated the legs and wings of the carcass and then applied elasticised string around the carcass before tossing it onto a moving conveyor belt. The court was satisfied on the evidence that there was a sufficient causal link between the work on the trussing line and the exacerbation of her symptoms. The question then was whether this manipulation was carried out as part of any manual handling operation.[17]

Clearly moving the chicken carcass to the workbench and throwing the trussed carcass onto the conveyor belt could be categorised manual handling since she was transporting or supporting a load by hand or by bodily force.

However, crucially, there was no suggestion that Mrs Hughes sustained any injury or exacerbation of symptoms as a result of either of these steps. Her complaint was focused on the repetitive movements of the wrists, hands and fingers in the manipulation of the chicken carcass, by tucking in the legs and wings and tying it with elasticated trussing string.

While this manipulation was being performed, there was no transporting or supporting of a load. It followed from this that the exacerbation of symptoms experienced by Margaret Hughes as a result of manipulating

---

[14] An unreported decision of HHJ Serota QC sitting in the Milton Keynes CC. The injury occurred through repeated handling of moderate loads over time such as boxes of stationery.

[15] [2003] EWCA Civ 771. The Court of Appeal confirmed a causative breach of the Regulations in a case in which the claimant suffered no specific accident but developed a back condition over time through moving patients without any of the appropriate equipment, assistance and training.

[16] [2007] CSIH 32.

[17] See reg 2(1) below.

the chicken carcasses on the trussing line was not caused by a breach of reg 4(1)(a) of the 1992 Regulations. She lost.

9.3.3 What is Manual Handling?

The key words in reg 2(1) of Manual Handling Operations Regulations 1992 are 'any transporting or supporting of a load'. However it does not follow that because an employee is pushing or pulling or moving something that this necessarily involves the transporting or supporting of a load. A good example is *King v Carron Phoenix*.[18] King claimed damages for tennis elbow allegedly caused by repeated use of a spanner to tighten and loosen nuts. Lord Kingarth stated:

> 'In my view as a matter of ordinary language, and in the context of the regulations, although the pursuer was no doubt involved in pushing and pulling when working with the spanner, it could not be said that he was involved in the transporting or supporting of a load.'

The spanner was clearly a piece of work equipment. However the fact that a load is also work equipment does not immediately take it outside the Regulations. For example in *McIntosh v City of Edinburgh Council*[19] Lord McEwan held that a 'ladder' that had been used to gain access to a roof in need of repair could also be a 'load' when being handled and dismantled. Clearly that is correct.

9.3.4 Duties of Employers

The duties of employers are set out in reg 4 of Manual Handling Operations Regulations 1992. Regulation 4 (1)(a) requires that an employer shall, so far as is reasonably practicable, avoid the need for his employees to undertake[20] any manual handling operations at work which

---

18 An unreported decision of the Outer House of Lord Kingarth in the Court of Session on 26 January 1999.

19 [2003] SLT 827.

20 As Lady Smith held in *Napier v Scotsman Publications Ltd* [2004] ScotCS 134 the duty contained in reg 4(1)(a) of the 1992 Regulations only arises in circumstances where an employee needs to undertake a manual handling operation. Such circumstances may arise where an employee is specifically instructed to do so or where manual handling is the only way to carry out a required task. Napier was moving a pump with a colleague, Michael Keenan, and the lifting equipment they planned to use turned out to be unsuitable. Napier, a very experienced and senior employee, tried to move it by hand and was injured. Lady Smith held: 'The reason why the pump came to be manually handled was the exercise of choice on the part of the pursuer and Michael Keenan in circumstances where the manual handling of a substantial load was not the only option available to them. It was clear from the evidence that there were two other options available namely the use of the jib crane and going to speak to Mr Tulloch before attempting to move the pump, once the problem with the use of the pallet truck had arisen.'

involve a risk of their being injured.[21] That is the starting point. Only if that cannot be done does the employer move on to the next stage.

The burden on the employer is to avoid manual handling. There is no need to look at the rest of reg 4 at all unless the employer pleads and proves that it could not be avoided.

This was made clear by the Court of Appeal in *King v RCO Support Services Ltd & Yorkshire Traction Co Ltd*[22] On the day of the accident King arrived at work to find that the bus yard was covered in ice and was, therefore, slippery. In accordance with normal practice, he set about the task of spreading grit over the surface of the yard. That grit was kept in a pile within the yard and he made use of a sack barrow and a shovel. The yard was big and after two hours he had only managed to grit about two-thirds of it. He stopped gritting and then foolishly stepped off the gritted area and slipped on the untreated ice and was injured.

The Court of Appeal held that the task 'involved' the moving of a pile, or load, of grit from the place where it was kept so as to distribute it over the yard. That was the 'task' that he was performing and it came within the definition of what was a 'manual handling operation'.

On the basis that the employer did not put forward a defence that it was not reasonably practicable to avoid the risk, they held that the court had to presume that it was. If it was possible to avoid the task involving manual handling there had to be primary liability due to the breach of the regulations. The employer was liable as they were in breach of reg 4(1)(a).

It is important to remember that he was not 'manual handling' at the time he was injured. However he was 'involved' in a manual handling task and so was entitled to full protection under the Regulations.

That any task including *'elements'* of manual handling must be looked at as a whole is also shown clearly in *Wiles v Bedfordshire CC*.[23] A residential social worker employed by the council was injured when taking a disabled girl to the lavatory at a respite residential care home. She regularly looked after the girl who weighed 29.5kg and often took her to the lavatory on her own.

The girl used an electric wheelchair to move about, could weight bear for short periods but not walk. The girl's upper limbs were prone to occasional involuntary spasms. The social worker took the girl to the lavatory and lifted her out of her wheelchair so she could lean against a support bar whilst the social worker crouched down to remove her

---

21 See the flowchart at the end of this chapter for a simple explanation of the application of the Regulations.

22 [2001] ICR 608.

23 (2001) 6 CL 365.

underclothes. Whilst in this position the girl threw out her hands unexpectedly and fell backwards against the social worker causing her a back injury.

It was held that the 1992 Regulations applied and had been breached even though, at the precise time of the accident, Wiles was not actually manually handling the girl. The manual handling 'task' was taking the girl to the lavatory.

The conviction of *Julian Tyrrell Felt Roofing Ltd*[24] is a clear reminder where an employer must start. One of their employees was injured falling a distance of 4 metres from a ladder when he was carrying roofing materials. They were convicted of a breach of reg 4(1)(a) because manual handling had not been avoided and the risk had not been assessed. Elimination of manual handling whilst climbing ladders was said to be reasonably practicable. If there is such a risk and the employer does not eliminate it then the burden is on them to show that it was not reasonably practicable to avoid the need for the manual handling.

9.3.5 Reversing the Burden of Proof

As we have seen, the only time that it becomes necessary to move beyond reg 4(1)(a) of Manual Handling Operations Regulations 1992 is when the employer claims that it is not reasonably practicable to avoid the manual handling. The onus of proving what is reasonably practicable always rests with the defendant. The burden of proof accordingly is reversed; the employer must prove compliance. It is not for the employee to establish the breach. For decades, on the authority of the House of Lords, this has been accepted as the position.[25] If there was any doubt in *Bilton v Fastnet Highlands Ltd*[26] Lord Nimmo Smith confirmed that the same applies to health and safety regulations.

---

24 (unreported) 7 March 2001, Wantage Magistrates' Court.

25 The key case is *Nimmo v Alexander Cowan & Sons Ltd* [1968] AC 107. During the course of his employment Nimmo was unloading bales of pulp. He stood on one of the bales for the purpose of unloading others, fell due to the tipping of the bale, and was injured. He sued his employers, Alexander Cowan & Sons Ltd, for breach of statutory duty under the Factories Act 1961, s 29(1), which provides that there 'shall, so far as is reasonably practicable, be provided and maintained safe means of access' to working places and 'every such place shall, so far as is reasonably practicable, be made and kept safe for any person working there'. Nimmo did not allege that it was reasonably practicable for the respondents to make the working place safe. The House of Lords held that, on the true construction of s 29 and in view of the fact that a criminal offence was created, the onus of proving that it was not reasonably practicable to make the working place safer than it was lay on the employer; and accordingly that it was not necessary for the claimant to allege that it was reasonably practicable.

26 (1997) Times Law Report, 20 November.

The position is set out as a statutory requirement in reg 9 of the Merchant Shipping and Fishing Vessels (Manual Handling Operations) Regulations 1998:[27]

> 'In any proceedings for an offence under any of these Regulations consisting of a failure to comply with a duty or requirement to do something so far as is reasonably practicable, it shall be for the defendant to prove that it was not reasonably practicable to do more than was in fact done to satisfy the duty or requirement.'

The claimant does not have to plead that it was reasonably practicable to do have done something, although for tactical reasons that may be a choice the claimant makes. There are a number of cases that are worth looking at to see how the law works in practice and can be used to the claimant's benefit.

The leading case is *Davidson v Lothian and Borders Fire Board*[28] a decision of the Inner House (Appeal Court) of the Court of Session. It involved injury to a firefighter who was, with colleagues, engaged in putting up a large ladder during a drill.

Lord Macfadyen set out the requirements on the pursuer or claimant. In order to make a case under reg 4(1)(b)(ii) all that he was required to prove was that:

(i) he was engaged in a manual handling operation;

(ii) it gave rise to a risk of injury;

(iii) an event falling within the ambit of that risk occurred; and

(iv) actual injury was thereby caused.

If these matters were proven Davidson was entitled to succeed *unless* the defenders made out the statutory defence that they had taken appropriate steps to reduce the risk of injury to the lowest level reasonably practicable.

Davidson won because they could not satisfy that burden. It was inherent in the manual handling operation undertaken that it involved the carrying through of the drill in a manner that approximated to the way in which it would normally be carried out in operational conditions. They concluded that the postponement of the operation to await more favourable weather conditions could be regarded as an 'appropriate step' under reg 4(1)(b)(ii), as would a decision to abort because of a deterioration in weather conditions.

---

27 SI 1998/2857.

28 2003 SC Lothian R 750.

The first issue 'that he was engaged in a manual handling operation' is straightforward and what can be a 'manual handling operation' is covered earlier in this chapter.

The second issue that it gave rise to a risk of injury needs to be considered. How big a risk is needed? In *Anderson v Lothian Health Board*[29] Lord Macfadyen set the standard saying that 'injury need be no more than a foreseeable possibility'. That was accepted by the Court of Appeal in *Koonjul v Thameslink Healthcare Services NHS Trust*.[30] They clarified that by holding that the risk had to be a real risk that was a foreseeable possibility.

Issues three and four are self evident. A manual handling risk becomes a reality and causes injury. It really is as simple as that.

So the employer cannot avoid the risk. They must now turn to reg 4(1)(b) that goes on to look at what an employer must do where it is not reasonably practicable to avoid the need for his employees to undertake any manual handling operations at work that involve a risk of their being injured.

It starts with the requirement under reg 4(1)(b)(i) to make a suitable and sufficient assessment of all such manual handling operations to be undertaken by them, having regard to the factors set out in column 1 of Sch 1 to the Regulations. When carrying out the risk assessment the employer must consider the questions set out in the corresponding entries in column 2 of the Schedule. The schedules are reproduced at the end of this chapter.

Regulation 4(2) imposes an absolute duty on the employer to review any assessment made if there is reason to suspect that it is no longer valid or there has been a significant change in the manual handling operations to which it relates. Where, as a result of a review, changes to an assessment are required, the employer has an absolute duty to make them. The most obvious reason to review an assessment would be someone suffering injury as that would automatically suggest that the original assessment had not covered everything fully.

Following the risk assessment the employer is required to comply with reg 4(1)(b)(ii) by taking appropriate steps to reduce the risk of injury to from the manual handling operations to the lowest level reasonably practicable. In most cases this duty is only complied with fully when the risk of injury is reduced to a level where it is virtually eliminated. For a residual risk to be one still needing reduction, injury need not be a probability.

---

[29] SCLR 1086.

[30] (2000) Times Law Report, 19 May, CA (Hale LJ, Sir Christopher Staughton) .

Here are three examples of reasonably practicable steps not taken which were enough to establish a breach:

- the failure to ensure that an experienced and trained worker attended refresher manual handling training;[31]
- the failure to show a training video to the claimant designed to train employees not to react instinctively by twisting around if he was called from behind by a fellow worker as he carrying a load;[32]
- the failure to impress on the mind of an employee not to walk on uneven surfaces.[33] The Court of Appeal held that this was so although it might seem to be training for something obvious.

Next the employer must comply with reg 4(1)(b)(iii) by taking appropriate steps to provide any of the employees undertaking the manual handling operation with general indications and, where it is reasonably practicable to do so, precise information on the weight of each load and which is the heaviest side of any load if its centre of gravity is not positioned centrally.

In *Swain v Denso Marston Ltd*[34] failure to provide this sort of information alone gave rise to full liability. In *McBeath v Halliday*[35] Lord Macfadyen held that the failure to issue manufacturer's instructions materially contributed to an accident to an electrician who was injured fitting electrical wiring to a floodlighting column.

9.3.6 Additional Duty on Employers

On 27 September 2002 the Health and Safety (Miscellaneous Amendments) Regulations 2002[36] brought in an amendment to reg 4 of the Manual Handling Operations Regulations 1992 by adding the following subparagraph:

> '(3) In determining for the purposes of this regulation whether manual handling operations at work involve a risk of injury and in determining the appropriate steps to reduce that risk regard shall be had in particular to -
> (a) the physical suitability of the employee to carry out the operations;
> (b) the clothing, footwear or other personal effects he is wearing;
> (c) his knowledge and training;
> (d) the results of any relevant risk assessment carried out pursuant to regulation 3[37] of the Management of Health and Safety at Work Regulations 1999;

31 *Walsh v TNT UK Ltd* 2006 CSOH 149.
32 *O'Neill V DSG Retail Ltd* [2002] EWCA Civ 1139.
33 *Smith v S Notaro Ltd* [2006] EWCA Civ 775.
34 (2000) PIQR P129.
35 (2000) 6 CL 731.
36 SI 2002/2174.
37 This imposes a duty on every employer to make a suitable and sufficient assessment

(e) whether the employee is within a group of employees identified by that assessment as being especially at risk; and

(f) the results of any health surveillance provided pursuant to regulation 6[38] of the Management of Health and Safety Regulations 1999.'

There are no reported cases based on any breach of these new requirements. Their aim is clearly to ensure that employers recognise that risk assessments under the 'management' regulations must be acted upon and the regulations should be looked at together.

9.3.7 Duty of Employees

The main line of defence to these cases is reg 5 of the Manual Handling Operations Regulations 1992 that imposes an absolute duty on all employees to make 'full and proper use' of any system of work provided for their use by the employer when the employer is complying with the requirements of reg 4(1)(b)(ii).

In other words if the employer is unable to avoid the manual handling and following a risk assessment implements actions to reduce the risk of injury to an acceptably low level the employee must use them. For example if, following a risk assessment, an employee is given a trolley to use they should use it.

In *Perry v Comedy Store Ltd CC*[39] Jacqueline Perry moved two boxes with a combined weight of 26kg injuring her back. The judge held that the Comedy Store had put into place a safe system of work under which there was no requirement for Perry to undertake such manual handling operations. The system involved a caretaker moving such loads for her. Perry was held to have breached her duty as an employee under reg 5 and she lost.

9.4 Further Regulations

The Merchant Shipping and Fishing Vessels (Manual Handling Operations) Regulations 1998[40] came into force on 31 December 1998. They were required because reg 3 of the Manual Handling Operations Regulations 1992 excluded their application to or in relation to the master

---

of the risks to the health and safety of his employees to which they are exposed whilst they are at work; and the risks to the health and safety of persons not in his employment arising out of or in connection with the conduct by him of his undertaking.

[38] This imposes a duty on every employer to ensure that his employees are provided with such health surveillance as is appropriate having regard to the risks to their health and safety which are identified by the assessment carried out under reg 3.

[39] An unreported decision of Recorder Eady in the Central London County Court on 24 May 2006.

[40] SI 1998/2857.

or crew of a sea-going ship or to the employer of such persons in respect of the normal ship-board activities of a ship's crew under the direction of the master.'

The Regulations apply to all UK ships, other than public service vessels or vessels engaged in search and rescue which are excluded in reg 3(1) or when the Manual Handling Operations Regulations 1992 or reg 3(3) of the Manual Handling Operations Regulations (Northern Ireland) 1992. Non UK ships in UK waters are subject to regs 3(2) and 10-13 covering inspection and detention.

The approach is the same as in the earlier regulations with an obligation under reg 5(1) of the Manual Handling Operations Regulations 1992 placed on the employer to avoid, so far as is reasonably practicable, the need for any manual handling of a load that would involve a health and safety risk to the worker. Appropriate measures to avoid the need for manual handling of loads that involve a risk of workers being injured specifically includes considering mechanical equipment. If avoidance is not reasonably practicable the employer is under an absolute duty to carry out an assessment having regard to specified factors and consider specified questions in relation to those factors set out in a schedule.

There are matching duties to take appropriate steps to reduce the risk of injury to workers to the lowest level that is reasonably practicable (reg 5(2)(b)); take steps to provide the worker with precise information on the weight and centre of gravity where it is practicable to so (reg 5(2)(c)), and provide workers who will be involved in a manual handling operation with proper training and information (reg 5(2)(d)).

Under reg 4, obligations on the employer can be extended to another person, such as the ship's master, if the employer is not in control of the matter because he does not have responsibility for the operation of the ship. An obligation under reg 6 is placed on the worker to make full use of any system of work provided by the employer to reduce the risk to the lowest level that is reasonably practicable.

The Regulations impose criminal penalties for breaches. Under reg 7(1) contravention of the requirement on the employer (or, if appropriate, another person in control of the matter) is an offence with a maximum penalty of £5,000 and under reg 7(2) contravention of the requirement on the worker is an offence with a maximum penalty of £500. Under reg 8 provision is made for offences by a corporate body or a Scottish partnership and under reg 9 the burden of showing that failure to comply with a duty in these Regulations was not reasonably practicable is on the defendant. That is of course the situation in a civil case in any event.

Under the Merchant Shipping And Fishing Vessels (Manual Handling Operations) Regulations 1998 the factors to which the employer must have regard and questions he must consider when making an assessment of manual handling operations or providing instruction for workers are set out. They are very similar to those in Sch I to the 1992 Regulations but with specific thought to the risks that those on board ship face such as uneven, slippery or unstable deck surfaces and variations in level of deck surfaces such as door sills or work surfaces.

APPENDIX 1 FLOW CHART

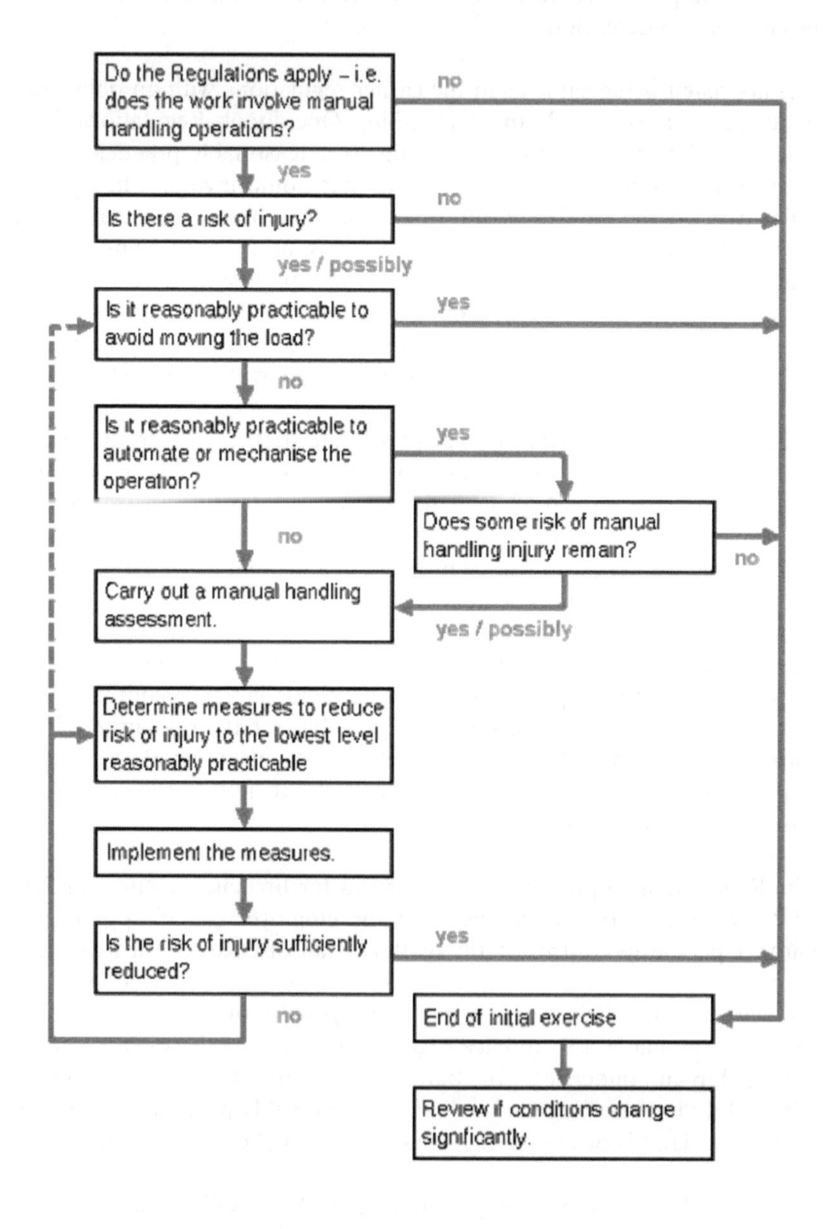

APPENDIX 2 FACTORS TO WHICH EMPLOYER MUST HAVE REGARD WHEN MAKING ASSESSMENT OF MANUAL HANDLING OPERATIONS

Regulation 4(1)(b)(i) of Manual Handling Operations Regulations 1992 requires the employer to make a suitable and sufficient assessment of all such manual handling operations to be undertaken by them, having regard to the factors set out in column 1 of Sch 1 to the Regulations. When carrying out the risk assessment the employer must consider the questions set out in the corresponding entries in column 2 of the Schedule.

Column 1 *Factors*	Column 2 *Questions*
1. The tasks	Do they involve: — holding or manipulating loads at distance from trunk? — unsatisfactory bodily movement or posture, especially: — twisting the trunk? — stooping? — reaching upwards? — excessive movement of loads, especially: — excessive lifting or lowering distances? — excessive carrying distances? — excessive pushing or pulling of loads? — risk of sudden movement of loads? — frequent or prolonged physical effort? — insufficient rest or recovery periods? — a rate of work imposed by a process?
2. The loads	**Are they:** — heavy? — bulky or unwieldy? — difficult to grasp? — unstable, or with contents likely to shift? — sharp, hot or otherwise potentially damaging?

Column 1 *Factors*	Column 2 *Questions*
3. The working environment	**Are there:** — space constraints preventing good posture? — uneven, slippery or unstable floors? — variations in level of floors or work surfaces? — extremes of temperature or humidity? — conditions causing ventilation problems or gusts of wind? — poor lighting conditions?
4. Individual capability	**Does the job:** — require unusual strength, height, etc? — create a hazard to those who might reasonably be considered to be pregnant or to have a health problem? — require special information or training for its safe performance?
5. Other factors	Is movement or posture hindered by personal protective equipment or by clothing?

CHAPTER 10

ACCIDENTS ON CONSTRUCTION SITES

10.1 INTRODUCTION

Construction sites are dangerous places. The construction industry employs approximately 7% of the UK workforce but accounts for 25% of fatal accidents and 16% of the major accidents.[1]

This chapter will deal with common problems that arise in pursuing a claim for injuries arising from accidents on construction sites. It will focus on where duties lie to prevent accidents, and how to select the correct defendant, which is one of the main difficulties facing the claimant in a construction case. The problem can become more acute, in establishing an entitlement to damages, when one or more of the prospective defendants has gone out of business or where there is no employers' liability insurance in place.

It will be seen that the key to determining who to sue is the question of control of the work.

10.2 MAIN SOURCES OF LAW

The relevant law governing health and safety on construction sites is found in the Construction (Design and Management) Regulations 2007[2] ('CDM Regulations 2007'), which came in to force on 6 April 2007. These bring together the earlier Construction (Design and Management) Regulations 1994[3] ('CDM Regulations 1994') and the Construction (Health Safety and Welfare) Regulations 1996[4] ('CHSW Regulations 1996') and in part provides for stricter regulation than was previously in place.

---

1 Para 7.1 of the Explanatory Notes to the Construction (Design and Management) Regulations 2007, SI 2007/320.

2 SI 2007/320.

3 SI 1994/3140.

4 SI 1996/1592.

In addition the Work at Height Regulations 2005[5] came into force on 16 March 2005 and revoked part of the CHSW Regulations 1996 with a view to tightening the regulatory framework. The Regulations apply to all work places and not just construction sites. (For details about these Regulations see Chapter 5 on the Workplace Regulations.)

10.2.1 What is Construction?

In the vast majority of cases it will be readily apparent that the work being undertaken at the time of an accident involves the application of the regulatory framework. Building work is self evident but the ambit of the regulations is wider than that and is set out in reg 2(1) of the CDM Regulations 2007 and includes: renovation, redecoration or other maintenance (including high pressure water cleaning), demolition; the installation, commissioning, maintenance and repair or removal of services that are normally fixed within or to a building such as electricity, gas, telecommunications, and computers. This is a wide-ranging definition and goes well beyond putting up a building.

However, where a specific work shop, for example to deal with fabrication, is set up the Workplace Regulations will apply.

10.2.2 Construction and Design

CDM Regulations 1994 and Parts 2 and 3 of CDM Regulations 2007 impose a statutory scheme on those planning a construction project before the first sod is turned and throughout the construction.

The general scheme is that when setting out the plans health and safety must be at the forefront of the thinking and that those involved coordinate and produce, in effect, a joint scheme for the safe execution of the work.

The CDM Regulations 2007 differentiate between notifiable (to HSE) and non-notifiable projects. A project is notifiable if the work will, or is likely to, last for more than 30 days or involves more than 500 person days on site.[6] On these larger projects there are increased duties on the participants.

---

[5] SI 2005/735.

[6] One person on site for one day is one person day; 20 people on site for one day or 10 people on site for two days is 20 person days and so on.

10.2.3 The Participants

10.2.3.1 The Client

Under reg 2(1) of the CDM Regulations 2007 the client is the person who in the course or furtherance of a business seeks or accepts the services that may be used in carrying out a project or who carries out the project himself.

The client must check the competency of those he appoints, appoint a planning co-coordinator (on notifiable projects) and principal contractor, ensure suitable management and that the plan relating to health and safety during construction is accessible.

10.2.3.2 Planning Coordinators

In respect of a notifiable projects appointment of project co-coordinators is mandatory. The role is to carry out the tasks necessary for the client to comply with the Regulations and in particular to check the competency of the appointees of the client and produce the health and safety file for the project.

10.2.3.3 Designers

This may include the client and contractors. Again they are obliged to take health and safety matters on board and to consider the risks to the end (or ultimate) user of the building.

10.2.3.4 Principal Contractors

They are required to check the competency of subcontractors and to plan the construction phase in all its aspects and in doing so they are obliged to consult the workforce. Other contractors must ensure that the works they carry out are planned, managed and monitored.

10.2.4 Civil Liability

Under CDM Regulations 1994, reg 10, civil liability was imposed on the client in respect of the duty to ensure, so far as reasonably practicable, that a safety plan was in place before the start of the construction phase. That general liability disappears under CDM Regulations 2007 and is replaced by the more specific duties set out above.

The scheme of reg 45 CDM Regulations 2007 is to broaden civil liability in respect of the planning stage to bring in the client and the principal and other contractors but only to a limited extent, as follows:

Regulation 45 allows an employee of the participants to bring a claim for breach of any of the statutory duties by the employer. That ambit of liability is extended in only limited circumstances as follows:

- the client's duty under Sch 2 which contains provisions for welfare facilities – sanitary and washing connivances, drinking water, changing rooms and lockers, and rest facilities: reg 9(1)(b);
- the duty of contractors to prevent unauthorised access to the site: reg 13(6);
- the contractor's duties re Sch 2: reg 13(7);
- the client's duty to ensure that the principal contractor has complied with regs 22(1) and 23 (health and safety plan) before work commences: reg 1(1);
- the principal contractor's duty to ensure that Sch 2 is complied with throughout the construction phase: reg 22(1)(c);
- the principal contractors duty to take reasonable steps to prevent unauthorised access: reg22(1)(l).

The most important aspect, and which is a continuation of the duties under CHSW Regulations 1996, is found in reg 25(1), (2) and (4). This gives rise to civil liability for contractors and every person having control of the way in which any construction work is carried out in respect of duties under regs 26–44.

10.2.5 Impact of Preconstruction Planning Duties on Civil Liability

The planning phase relating to health and safety is likely to produce generalised documentation setting out the duties of those involved by reference to the statutory duties. However, most accidents are caused by system or human failings in the detailed application of the laid-out criteria. In those circumstances the main search for causative breach of statutory duty is likely to be directed at CDM Regulations 2007, regs 26-44, which replace CHSW Regulations 1996. Liability for breach (see below) is based upon control of how, or the conditions in which, work is undertaken.

The duties contained in Sch 2 may be useful in cases of disease, especially dermatitis, if the client is to be a party. However, if the client plans the facilities and they are not put in place or maintained then the question of control will be the focus of the investigation.

10.3 Construction Phase Duties

Given that consideration of the CDM Regulations 2007 and the 1996 Health and Safety provisions will both need to be used for some time to come it is useful to compare the provisions. The first matter to note is that the definition of those having responsibilities has changed:

10.3.1 CHSW Regulations 1996

'4.—Persons upon whom duties are imposed by these Regulations

(1) Subject to paragraph (5), it shall be the duty of every employer whose employees are carrying out construction work and every self-employed person carrying out construction work to comply with the provisions of these Regulations insofar as they affect him or any person at work under his control or relate to matters which are within his control.

(2) It shall be the duty of every person (other than a person having a duty under paragraph (1) or (3)) who controls the way in which any construction work is carried out by a person at work to comply with the provisions of these Regulations insofar as they relate to matters which are within his control.

(3) Subject to paragraph (5), it shall be the duty of every employee carrying out construction work to comply with the requirements of these Regulations insofar as they relate to the performance of or the refraining from an act by him.

(4) It shall be the duty of every person at work—
(a) as regards any duty or requirement imposed on any other person under these Regulations, to co-operate with that person so far as is necessary to enable that duty or requirement to be performed or complied with; and
(b) where working under the control of another person, to report to that person any defect which he is aware may endanger the health or safety of himself or another person.
(5) This regulation shall not apply to regulations 22 and 29(2), which expressly say on whom the duties are imposed.'

10.3.2 CDM Regulations 2007

'25. Application of Regulations 26 to 44

(1) Every contractor carrying out construction work shall comply with the requirements of regulations 26 to 44 insofar as they affect him or any person carrying out construction work under his control or relate to matters within his control.

(2) Every person (other than a contractor carrying out construction work) who controls the way in which any construction work is carried out by a

person at work shall comply with the requirements of regulations 26 to 44 insofar as they relate to matters which are within his control.

(3) Every person at work on construction work under the control of another person shall report to that person any defect which he is aware may endanger the health and safety of himself or another person.

(4) Paragraphs (1) and (2) shall not apply to regulation 33, which expressly says on whom the duties in that regulation are imposed.'

The allocation of responsibility has been streamlined into a more sensible and straightforward scheme based upon control and those affected as a result of the control or extent of it.

The following table shows the changes made to the Regulations and the destination of the old to the new positions in the Regulations. In addition it notes the level of duty imposed.

Duty	1996 Regs	2007 Regs	Comments
Safe access, egress and place of work	5	26	Regulation 5(4) which provided an, that the level of duty was so far as **practicable**, to ensure the safety of a person who was making the place safe has been removed. The duty now is to take all reasonably practicable steps; which is a lowering of the standard required.
Good order and security	26	27	No change. General duty is reasonable practicality. Note the absolute duty re nails in reg 26(3) is extended in reg 27(3) to include any similar sharp object.
Falls, fragile material, falling objects	6–8		See Work at Height Regulations 2005.
Stability of structures	9	28	The duty remains one of all practicable steps.
Demolition and dismantling	10	29	The duty remains one of all practicable steps. Work must be carried out under competent supervision.
Explosives	11	30	No change. Reasonably practicable.

Duty	1996 Regs	2007 Regs	Comments
Excavations	12	31	No change. All practicable steps.
Cofferdams and cassions	13	32	All practicable steps. 32 (1) (a)) introduces a new duty to provide appropriate means of escape if water or materials enter the structure.
Preventing drowning	14	35	No change. Reasonably practicable.
Traffic routes	15	36	No change. Reasonably practicable.
Doors and gates	16	Deleted	Absolute duties re safety aspects.
Vehicles	17	37	Suitable and sufficient safeguards in respect of unintended movement, warnings to be given in respect of movement of vehicles, prohibition of travel unless in a safe place, prevention against falling over the edge of an excavation.
Prevention of risks from fires	18	38	Reasonably practicable.
Emergency routes and exits and emergency procedures	19, 20	39, 40	Emergency procedures which are suitable and sufficient must be in place re foreseeable emergencies. Emergency routes must be suitable and sufficient. The duties to keep escape routes free from obstruction and provided with emergency lighting are absolute.
Detection and fighting fires	21	41	No change. Suitably located and suitable and sufficient equipment and detectors. Detectors must be maintained – absolute duty.
Welfare facilities	22	Sch 2	Various duties re changing rooms, washing facilities, drinking water. etc. Now includes separate facilities for women.

10.4 CONTROL

Construction sites will often confront the investigator of a claim with a number of potential defendants. There will be a main contractor and sub-contractors and some of the sub-contracted work will in turn be further sub-contracted. It is not unusual for insurers to try to pass the parcel of blame between different parties on the basis of who was responsible for the accident.

In addition arguments may be raised that Party A has agreed to indemnify Party B and they should therefore deal with the claim. In practical terms if the situation is accepted by B then there is no problem in dealing with that party directly – but once they start raising issues as to whether there is an indemnity or the extent of it, then go back to A if that was the party with control and leave them to sort out the contractual arrangements.

Responsibility for breach of the health and safety provisions is based upon control of the situation giving rise to the injury. It is a key concept in respect of selecting the correct defendant.

The word 'control' is a simple one and involves assessment of the facts in the individual case. The question of control can be illustrated by considering three cases.

10.4.1 McCook v Lobo[7]

The claimant used a ladder to gain access to a ceiling and inevitably, because it was unfooted, the ladder slipped and he fell. The difficulty in the case arose because the employer, Hedley, had no insurance or monies to pay the judgment. The claimant therefore sought to impose liability upon the other defendant who owned the building being renovated for business purposes and who employed Hedley to carry out the work. (A client under CDM Regulations 1994 or CDM Regulations 2007.)

The court rejected the propositions that the owner employed the claimant and a claim under the Occupiers' Liability Act 1957 on the basis that Hedley was a competent independent contractor. In addition it was found that although the defendant visited the site and dealt with some matters relating to health and safety it had no control over how this particular task was carried out.

The claimant sought to establish liability under the admitted breach of reg 10 of CDM Regulations 1994. There was no plan in place; however, the Court of Appeal dismissed the appeal for the following reasons:

---

7 [2002] EWCA Civ 1760.

- the plan would not have dealt specifically with the event that occurred;
- the defendant had no control over how it was to be carried out – that was left to the employer who disregarded elementary safety precautions;
- in those circumstances the breach was not causative of the injury suffered by the claimant.

The difficulty facing the claimant in such a case is a very real one. Hale LJ stated that liability depends upon control, which is a question of fact, and in this case the client did not have control; this had been passed to the reputable independent contractor.

One point not raised in the case is whether a client can be said to have engaged a reputable contractor without taking the step of ensuring that he had employer's liability insurance. If that was taken as evidence of lack of repute, which is a question of fact, then the claim might have succeeded at common law.

McCook is not an untypical example of a practicable problem that frequently occurs in construction cases. It might have been different in that case if there had been an agreement between the defendants not to use scaffolding on a job in order to save money. In such circumstances the owner of the premises, who would have been a client under CDM Regulations 2007, could be said to have control of the way in which the work was carried out.

10.4.2 Hood v Mitie Services[8]

The Post Office subcontracted its maintenance work, which fell within the meaning of construction, to Mitie, who carried out work on various sites as and when required. Work had to be carried out on an asbestos roof and, whilst doing this, an employee of Mitie fell through a skylight. The work was to be carried out using ladders, which would mean that the men were not just using them for access but working from them. It was clearly an unsafe way of working.

Mitie, having settled the claim by the claimant, then sought indemnity or contribution from the Post Office.

Again there was no plan under CDM Regulations 1994, reg 10 either specific to the task or the general way in which work at height was to be carried out. Judge Playford QC rejected the argument on the basis that the regulation could not create actual control of how the work was done.

8 [2005] All ER (D) 11 (Jul).

It is unlikely that the 2007 Regulations alter the position – because the Post Office would never be in control of the work being done and that would be the position whether or not the necessary work had been done on the planning stage.

10.4.3 King v Farmer[9]

This unreported decision deserves consideration. The claimant was a subcontracted painter and decorator. The defendant was the main contractor and had arranged for scaffolding to be put up at the gable end of a property. For security reasons no ladder was kept in place to the first stage. The claimant arrived alone one morning, put up his own ladder and went up it with a view to securing it when he got to the top.

The judge decided that the defendant did not have control. The interpretation of the Regulations was the ultimate in the literal approach and is very probably wrong. Regulation 4 of the CHSW Regulations 1996 provides that those who control the way in which construction work is carried out by a person **at** work the judge reasoned that the claimant had not started work until he had ascended the ladder and started to use his paint brush. Therefore, the defendant was not in control at the time the claimant went up the ladder. The case was pleaded on the basis of safe access to work (reg 5) and in those circumstances '**at** work' should have been taken in the broader sense of being on site – after all, the purpose of the Regulations is to ensure safety once the person arrives **at** work. If the decision is correct then it would render reg 5 in terms of access and egress pointless. (See also *PRP Architects v Precious Reid*[10] for a detailed discussion of the meaning of 'at work' and where the judge at first instance was held to be correct that using a lift to leave the building was 'at work'.)

10.5 SPECIFIC DUTIES

10.5.1 Safe Place of Work

Regulation 26 of the CDM Regulations 2007 provides that so far as it is reasonably practicable access to and egress for any place of work should be suitable and sufficient; and that the place of work, so far as is reasonably practicable, shall be made and kept safe and without risk to health to any person at work. In addition any place that it is not so maintained requires suitable and sufficient steps to be taken to prevent access or egress.

A common problem that arises is in respect of a situation where there is more than one way of getting to or from a place of work. The defendant

---

9 [2004] EWHC B2 (QB).

10 [2006] EWCA Civ 1110.

will raise the argument that the claimant chose the unsafe route. The answer lies in the duty to take suitable and sufficient access to prevent access to a route which is unsafe – if that has not been done then there is a breach of statutory duty and the issue is really one of contributory negligence.

The following matters need to be taken into account in assessing contribution:

(a) Why was the claimant using route A rather than route B?

(b) If route A was the dangerous route, was it commonly used by other persons and was that known to supervisory staff?

(c) The extent of the problem on route B.

(d) How long it had been in that state.

(e) The time of day may be important as the courts will be less critical of a worker towards the end of the shift.

(f) If the route is said to be dangerous then were sufficient steps taken by the employer to prevent its use as a means of access or egress? In some cases it may be sufficient to tell employees not to use it and in others barring the way physically may be necessary.

10.5.2 Training and Experience

A common argument raised is that the claimant was an experienced worker and that therefore it was acceptable for the employer to expose him to some risk or that he should have known what he was doing was unsafe. In those circumstances it is said that the claimant has been the author of his own misfortune or that there is a very substantial degree of contributory negligence.

This notion is all too common and fails to approach the matter from the correct analytical aspect in that it looks at what the claimant did.

The first step is to consider whether there has been a breach of statutory duty by the employer. If there is a breach that has a causative effect then you go on to consider the respective contributions.

Boyle v Kodak[11] is authority for the following:

- Both employee and employer have duties under the regulations.

---

[11] [1969] 2 All ER 439, HL.

- The action of the employee is not decisive – one has to consider whether the employer has failed to perform its duties, and has this contributed to the accident.

- The employer can only escape liability if the sole cause of the accident is that of the employee.

- In order to show that the sole cause was that of the claimant the employer must prove that he took all reasonably practicable steps to prevent it.

The case makes clear that the employer cannot just leave the employee to get on with the job.

The matter was more recently considered by the Court of appeal in *Anderson v Newham College of Further Education*[12] where Sedley LJ emphasised the high standard of proof required to shift the whole blame to the claimant.

In more modern times the question of refreshing knowledge and emphasising the need for safety has been amplified as part of taking all practicable steps is to ensure training and education on a life-long basis. In this respect reference can be made to the HSE Approved Code of Practice issued under the CDM Regulations 2007 which sets out a plan for the continuing education of workers and encourages regular refreshers by way of tool box talks. While there is no civil liability for breach, the matters set out are statements of good practice and a failure to follow them may support an argument that the employer has failed to take all reasonably practicable steps to prevent the accident or at least defeat or contain an argument of contributory negligence. The importance of training, to prevent accidents, is illustrated in the manual handling cases of *Walsh v TNT UK Ltd*[13] and *O'Neill V DSG Retail Ltd.*[14]

The Management of Health and Safety at Work Regulations 1999[15] provide for risk assessments to be carried out by the employer and for these to be communicated to the employee – breach of which does give rise to civil liability.

In *General Cleaning Co v Christmas*[16] Lord Reid, speaking of the common law duty expressed the high level of the employer's duty:

> 'The question then is whether it is the duty of the appellants to instruct their servants what precautions they ought to take, and to take reasonable steps to

---

12 [2005] EWCA 505.
13 [2006] CSOH 149.
14 [2002] EWCA Civ 1139.
15 SI 1999/3242.
16 [1952] 2 All ER 1110.

see that those instructions are carried out. On that matter the appellants say that their men are skilled men who are well aware of the dangers involved and as well able as the appellants to devise and take any necessary precautions. That may be so, but, in my opinion, it is not a sufficient answer. Where the problem varies from job to job it may be reasonable to leave a great deal to the man in charge, but the danger in this case is one which is constantly found and it calls for a system to meet it. Where a practice of ignoring an obvious danger has grown up I do not think that it is reasonable to expect an individual workman to take the initiative in devising and using precautions. It is the duty of the employer to consider the situation, to devise a suitable system, to instruct his men what they must do, and to supply any implements that may be required …'

There was a further statement by Lord Oaksey:

> 'It is the duty of an employer to give such general safety instructions as a reasonably careful employer who has considered the problem presented by the work would give to his workmen. It is, I think, well known to employers, and there is evidence in this case that it was well known to the appellants, that their workpeople are very frequently, if not habitually, careless about the risks which their work may involve. It is, in my opinion, for that very reason that the common law demands that employers should take reasonable care to lay down a reasonably safe system of work. Employers are not exempted from this duty by the fact that their men are experienced and might, if they were in the position of an employer, be able to lay down a reasonably safe system of work themselves. Workmen are not in the position of employers. Their duties are not performed in the calm atmosphere of a board room with the advice of experts. They have to make their decisions on narrow window sills and other places of danger, and in circumstances in which the dangers are obscured by repetition.'

The common law duty of reasonable care in respect of safe system of work is a strongly expressed one. The regulatory scheme focuses on risk assessments and training has reinforced the high duty placed on the employer. But the above quotations show how strongly the courts have asserted the duty and have long recognised the level of responsibility placed on the employer in construction and other cases. In this respect, and although the common law tends to be submerged by the regulations, it is still important for advocates to put the common law duty before the trial judge to ensure that the Regulations are not seen as a new idea, or gold standard, imported from Europe.

PART C

EFFECTIVE CASE PREPARATION AND PRESENTATION

CHAPTER 11

INVESTIGATION GUIDELINES

This chapter examines how to approach the investigation of claims in a way that maximises the prospects of a successful outcome to litigation.

11.1 ADVISER INVESTIGATIONS

There is no magic formula to the proper investigation of claims, but there are themes and methods of working that can assist in setting an agenda to work from. However, it is essential not to be limited by this approach and think beyond these guidelines where appropriate.

Even the most seemingly straightforward case needs to be approached with care so that the facts of the matter are established (and, where possible, verified) and that this is done at an early stage.

The approach should be to make progress on liability issues as quickly and as accurately as possible. One tool that can be used is to picture the scene in court and how the case will be presented – and to conduct the investigation on the basis that the case will be in court the next day.

Accuracy is of central importance as the letter of claim will set out, albeit briefly, the circumstances in which the accident occurred. If the matter is litigated the first few paragraphs of the particulars of claim/statement of the case will provide the description, given by the claimant, of how the injuries arose. Consistency is the hallmark of credibility or reliability of the claimant's evidence. An error in setting out the case in the letter of claim or particulars of claim may rebound against the claimant if changes have to be made later down the line.

11.2 BASIC GUIDELINES

There are two sets of three letters, or trios, that can usefully be utilised as aids to the general approach to the investigation process.

11.2.1 The 3 WWWs

- What happened?

- Why did it happen?

- What are the legal consequences?

This trio provides a useful analytical framework within which to approach the investigation of a claim. In cases where there is a breach of an absolute duty the 'Why did it happen?' may become of less importance and in some cases may, once the facts are established, become of little relevance. In other cases where the duty is not strict, why an event occurred maybe of considerable importance.

The framework is not to be seen as three separate steps in investigation but rather as a circle in the life of the case; because in order to understand what is relevant to the first two points it is necessary to have an understanding of the detailed legislative provisions that are set out in this book.

Specific examples of how to work through the process are featured later in this chapter.

In establishing the 'what and why' of the matter the first step is to talk to the claimant and obtain the information that is needed to establish the basic facts of the matter. In doing so you may be fortunate in having a client who is able to provide all the information that you need in one go – but often you will need to exercise patience and give the claimant time to think about matters. However, even with an articulate client it will be wise to ensure that the information is correct. A questionnaire may be useful so that the claimant can collect the information to reduce guesswork in respect of time, distances, measurements etc.

11.2.2 The 3 PPPs

- Previous accidents.
- Previous complaints.
- Post-action remedial action.

This second trio of strands of investigation is a process that will need to be considered very frequently and may make the difference between success and failure.

In cases where an absolute duty is not imposed foreseeability is likely to be an issue. It can be readily understood that previous accidents and complaints can be of importance. In addition close calls or near misses in the same or similar circumstances need to be looked into as evidence of foreseeability.

When a claimant says that he has complained about a situation, prior to the accident, it is necessary to drill into this and find out – when the complaint was made, to whom it was made and what was the reaction to it. The same routine applies to collection of evidence from both the claimant and all witnesses on this issue.

In some cases steps will be taken after the accident to remove a danger or to prevent an occurrence of an event. In such circumstances the defendant's employer may say the extra measures were taken as a counsel of perfection and that they could not be taken before because the danger was not foreseeable. This line of defence becomes less sustainable if there is evidence that there have been previous problems or complaints and the production of evidence will narrow or exclude the potency of such a line of argument.

11.3 THE ACCIDENT REPORT

The letter of claim will refer to the pre-action protocols on the question of disclosure. At the early stage of investigation of the claim the most important documents relate to the reporting of the accident and the contents need to be checked and verified by the claimant. If possible get sight of the documents before the first letter is sent. Often, if there is going to be an acutely disputed case it will be over how the accident happened. In such cases the description of the accident given by the claimant will be at odds with what is in the accident report.

Where this arises then it is necessary to go through the accident report with the claimant to ascertain:

Did he say what is recorded?

Under what circumstances was the report completed? A report completed immediately after an accident, particularly a serious accident, if the claimant is in pain or receiving treatment may be given less weight than one made when the claimant has had the opportunity to recover.

It may be the case that what the claimant has said has been paraphrased and inadvertently, or even deliberately, misdescribed by the employer.

In some cases the employer may have obtained details of how the accident occurred from someone else who has got it wrong.

11.4 SPECIFIC AREAS OF INVESTIGATION

11.4.1 Slip, Trips and Falls

The investigation of this category of accidents needs to produce a statement by the claimant that gives a clear and accurate description of the environment in which the injury occurred. The initial phone call or form completed by the claimant may be as brief as saying: 'I slipped on a floor'. Clearly, this is not going to carry the case forward and this brief narrative will need to be expanded considerably.

The following information needs to be obtained:

- Where was the claimant coming from and where was he going to?
- The time of day – the court can take into account that a worker may be less vigilant/alert towards the end of a shift.
- The lighting conditions.
- A description of the type of flooring, including colour and composition.
- A sketch plan showing the distance walked by the claimant before he fell. In particular the focus of the enquiry should be on the opportunity if any that the claimant had to observe the problem before the accident. This is important so that the level, if any, of contribution can be assessed.
- Was the claimant carrying, pushing or pulling anything, especially if this would have obscured the defect?
- Did anyone or anything reduce the visibility of the defect or substance?
- A description of the substance or defect that caused the fall. Here it is important to obtain dimensions.

It is very important to ask how did or how might the substance have got there. Remember that the employer is vicariously liable for breach of statutory duty as well as in negligence. If someone spills something on the floor they may have difficulty in showing that they took all reasonably practicable steps to prevent it happening.

11.4.2 Machinery

It is easy enough to visualise a drill, a chainsaw or a circular saw. However, manufacturing machinery, earth-moving machinery etc maybe more difficult for the claimant to describe or draw. However, the internet can become a very useful resource and if the claimant knows the manufacturer's details a photograph of the machinery will often be found on the internet. This can be brought up on the screen while the claimant is in the office and can simplify the most complicated of descriptions. It also means that the claimant can often point precisely to where the accident happened or identify the rogue part of the machinery.

If that is not available and in order to check the information provided by the claimant at an early stage the insurers should be asked to provide photographs that can then be checked over with the claimant.

11.4.3 PPE

Again the internet can be used to see what is available in order to illustrate what was provided and the investigate whether there was more or better equipment that would have further reduced any risk that the employee was exposed to.

11.4.4 Lifting Cases

The initial description you receive may just refer to lifting a heavy object! The claimant's description that a box or an object was heavy is not really very helpful in determining whether or not there has been a breach of the Manual Handling Regulations or a breach of the duty to provide a safe system of work.

In some cases the task will be relatively straightforward. The box or package, for example, may be labelled with the weight. However, there will be cases involving sheets of metal or lengths of wood where the information is not readily available. If you send off the letter of claim with the estimate made by the claimant the response may come back that the object has been weighed and is considerably less. Of course the employer's weight may be wrong in any event.

There can be simple ways of determining the issue. Some years ago (when consulting engineers were used much more) a trainee solicitor dealing with a claim involving lifting of bundles of newspapers asked his supervisor for permission to send an engineer in to weigh the bundles. The supervisor suggested buying a few copies and using the postroom scales to work out the weight of 50 copies in the bundle.

What it weighs may be difficult to determine. The following table may therefore be useful in reaching an accurate figure, to check what the client and the employer puts forward. It sets out the densities of various materials and if you can ascertain the volume of the object then the product of the two gives you the weight you are dealing with.

Density Table

	Grams/cubic centimetre
Steel	7.80
Cast iron	7.20

	Grams/cubic centimetre
Lead	11.40
Hard wood (Mahogany)	0.71
Soft wood (Pine etc)	0.56
Oak	0.77
MDF	0.6 to 0.8
Glass (Plate)	2.4
Water	1
Paper	0.7 to 1.2
Cardboard	0.69
Concrete	2.37

The figures for other materials can be found on various websites and there are also a variety of useful online converters to take you from metric to imperial measurements.

The weight of the object is the total volume multiplied by the density. Once you have the above information it becomes a simple task to calculate or estimate the overall weight.

Unfortunately, there will be objects that are so irregular in shape that you cannot estimate them; if all else fails then you, or the client, will have to weigh them.

11.5 SHADOW RISK ASSESSMENTS

The weight and dimensions of the object is only one part of the investigation. That information has to be applied to what was being done by the claimant in respect of moving or handling the object. In approaching this aspect it is essential to have the HSE *Guidance on Manual Handling* either in front of you or in your mind – so that essentially you are carrying out a risk assessment. In doing so you will create what the employer has failed to do or else check to see if the assessment has been done correctly to ensure that the risk of injury has been reduced to the lowest reasonably practicable level. In effect you can shadow what the employer did or did not do to establish whether there has been a causative breach of duty.

In this respect the HSE publication *Manual Handling Assessment Charts* are an invaluable tool that you should keep close to hand.

The *MHAC* considers lifting, carrying and team handling and the measures to take to identify risks and then reduce the overall level of risk of the task. The investigation of a lifting case can follow the factors identified in the booklet. The factors are common sense but the booklet provides a checklist for matters that you have to investigate.

In terms of lifting the headings are:

- load weight/frequency;
- hand distance from the lower back;
- vertical lift region;
- trunk twisting and sideways bending;
- postural constraints;
- grip on the load;
- floor surface;
- other environmental factors – heat, cold, wind etc.

When considering carrying, all the above are included, but the following are added:

- consideration of the position of the body's trunk to the load and whether it is asymmetrical;
- the distance to be carried;
- obstacles on the route.

Team lifting brings in additional complications and it is necessary to consider communication between the workers. The Guide suggests that lifting should be done on the count of '1, 2, 3'. So here you would want to look at who was appointed to call the lift on '3' and whether, for example, they began by saying clearly 'on 3'.

There are other factors such as the build of the person or persons doing the lifting. The claimant may have a history of back problems that the employer knew or did not know about.

However, if the above lists of factors are considered in producing the statement most if not all of the information needed to investigate and assess the prospects of success will be brought into account.

11.6 OFFICIAL INVESTIGATIONS

The Reporting of Injuries, Diseases and Dangerous Occurrences Regulations 1995[1] (RIDDOR) set out the duties for reporting to the HSE and local authorities. In serious cases there is a duty[2] to notify the authorities by the quickest practicable means.[3]

The most common document arising, from the duty, will be the prescribed form which must be sent off within 10 days of the accident which must be done if the injury causes absence from work.

The HSE and local authority have enforcement powers under the Health and Safety at Work etc Act 1974. In 2006/2007 there were 141,350 reported accidents and 241 fatal accidents reported but only 1400 prosecutions, approximately 1%. Therefore, in rare cases, involving death or serious injury, there will be useful documentation which may include witness statements and correspondence with the defendant and documents relating to any prosecution. The HSE/LA will require an order from the Court, to which they will consent, before releasing the documents. There are powers to issue prohibition[4] and/or improvement notices[5] – and these steps may have been taken before the accident and should be disclosed under the pre-action protocol by the defendant.

11.7 INQUESTS

It is invariably desirable to have the interests of the estate or dependents represented at an inquest. The death of an employee at work must be reported to the HSE in accordance with reg 4 of RIDDOR. In turn, s 8(3)(c) of the Coroners Act 1988 provides that the coroner must summon a jury in these circumstances. The ambit of the inquest is limited to considering when, where, and how the death occurred. The question of fault is not an issue and the coroner will not allow a trial of the question of responsibility.

Public funding is severely curtailed in the context of inquests, but the costs of representation (including cross-examination) for the purposes of gathering evidence are allowable on assessment in any successful civil claim.[6]

The fact-finding exercise of the inquest can be of significant assistance in the context of investigations and a positive verdict, while strictly only

---

1 SI 1995/3163.
2 See regs 3 and 10 RIDDOR SI 1995/3163.
3 SI 1995/3163, reg 3(1).
4 HSWA 1974, s 22.
5 HSWA 1974, s 21.
6 *Stewart v Medway NHS Trust* [2004] EWHC 9013 (Costs).

persuasive in any civil claim, will often influence the course of subsequent negotiations and, in default of agreement, the ultimate findings of the trial judge.

In practical terms contact should be made with the coroner's officer to check on the witnesses that it is intended to call; advisers should be alive to the need to ensure that all necessary witnesses are called and all relevant evidence is placed before the jury. Whilst the coroner presently retains liberal control over the conduct of the inquest, appropriately reasoned representations will normally receive a fair reception.

11.8 CONCLUSION

Remember that these are basic guidelines that are set out to give a grounding in the art of taking the statements that are the essential bedrocks of the establishment and progression of a personal injury claim. It is not a simple process, and often you will have to think outside the box to obtain all the relevant information to reach a successful conclusion.

CHAPTER 12

EVIDENCE

12.1 INTRODUCTION

Part B of this publication emphasises the closely regulated nature of health and safety in an employment context. This feature is of paramount importance when considering issues of case preparation and planning in all claims arising from accidents at work.

The injured employee already has a good head start when compared, for example, with a person, perhaps no less deserving, who is injured while engaging in a social or domestic activity. In most cases, it will be possible to frame the employee's claim with reference to one or more of the employer's statutory duties, which will typically attract strict liability if proved.

An accident at work claim is therefore most likely to fail on the basis of evidential weakness, in particular:

- insufficient evidence of breach of duty;
- insufficient evidence that an accident was caused or contributed to by the alleged breach of duty; and/or
- insufficient evidence that the injury, loss or damage complained of was caused or contributed to by the accident.

It is perhaps understandable, if certainly not excusable, that some advisers fall into the trap of leaving fuller investigations and evidential preparations to a late stage, lulled into a false sense of security that in many employers' liability cases such endeavours will not directly influence the third party's behaviour.

There are, however, a multitude of reasons why evidence should be collated and refined at the earliest opportunity. First and foremost, the weight attached to evidence is principally assessed with reference to cogency. The perceived cogency of evidence quickly diminishes over time, with delay at the outset of proceedings exerting a disproportionate effect.[1]

The quicker an adviser can marshal comprehensive and cogent evidence on all of the issues that are likely to prove relevant, the easier and more cost effective that exercise will be and the greater the weight that is likely to be attached to it.

Improved practices amongst some insurers and loss adjusters mean that early witness statements are now commonly obtained (in signed draft or CPR compliant format) at an early stage, contemporaneously with investigations during the protocol period.

Assuming early initial instructions are received and it is possible to initiate proceedings within a relatively short period of time, say between two and three months of an accident, it will nonetheless be at least another year before the filing and service of the injured party's lay witness evidence is likely to be directed by the court.[2]

If the evidence of the injured party, upon whom the burden will rest to at least establish a primary case, is prepared simply with reference to the procedural timetable (and thus may post date that of the employer's witnesses by a year or more), it is highly unlikely to attract the same weight.

While the above guidance is relevant in all cases, it has a growing resonance in this context. Employer's liability claims are becoming increasingly difficult to defend. Faced with the prospect of more effective presentation by employee advisers and a more consistent application of health and safety legislation by the courts, employer's liability insurers are faced with an increasingly stark choice – settle claims quickly so as to minimise the associated costs burden or look to resist or minimise claims by exposing evidential weaknesses. This chapter is intended to assist advisers to minimise the employer's opportunity to introduce uncertainty.

12.2 INITIAL STEPS

The importance of initial client contact cannot be understated. While the modern realities of acquisition mean that advisors will often represent

[1] Delay might properly be described as exerting an inverse, exponential effect upon the cogency of evidence. Trial judges are most likely to attach decisive weight to a period of a few days or a few weeks, for example, in the case of contemporaneous documentation, rather than to whether a witness was first asked to recall a relevant matter 18 or 24 months after an accident.

[2] Allowing up to six months between the accident and a protocol response, a further three months to obtain medical evidence and commence proceedings, between two and three months to obtain first directions and then a further 10 weeks for exchange based upon standard fast track time limits.

clients over a wide geographical area, face-to-face preliminaries should always be the rule rather than the exception where other constraints permit.

The ongoing effects of injury, anxiety about an unfamiliar legal process and funding issues will often weigh heavily on the injured person. It is often much easier to gain insight and offer an appropriate level of reassurance on a personal basis.

While important in every accident claim, this is perhaps more particularly so in the context of those arising at work. The quicker the client's full confidence is obtained, the quicker an accurate picture of the accident circumstances is likely to emerge. The adviser can then start to frame the case, at least on a preliminary basis, with reference to any potentially relevant health and safety requirements. The consultation can then focus upon more detailed issues such as the employer's general compliance and what documentary or lay evidence is likely to be available.

With modern word-processing and case management facilities, the traditional distinction between the client's initial 'proof of evidence' and 'witness statement' is becoming increasingly blurred. The initial consultation should not be too formulaic or closely circumscribed, but checklists or questionnaires have their place, if only to guard against the risk of obvious omissions.

The basic, essential information requirements in any accident at work claim are as follows:

- personal details: full name, date of birth and NI number;
- contact details: address and telephone number;
- employer details: address and business of undertaking;
- employment history: qualifications, experience and past jobs;
- employment details: job title and/or job description;
- medical details: GP and any relevant medical history.

The initial consultation must establish the scene. The date, time and exact location of the accident (and/or onset of symptoms) are obviously of central importance.

Details of any witnesses (their names and addresses, if known) and anyone to whom the accident was reported (or with whom it was later discussed) are also clearly important. If there is a shop steward or safety representative at the undertaking, appropriate details should be obtained.

The position and significance of any other person, property or equipment must be ascertained to provide a clear and complete picture of how the injured person came to suffer an injury. It is often helpful to invite or assist the client to draw a plan or diagram as appropriate.

If following the initial consultation there is any residual ambiguity as to how the accident occurred, prompt consideration should be given to arranging a site inspection and/or for photographs of the accident location and any relevant equipment to be taken, as discussed below. If the circumstances of the accident are not properly understood at the outset of proceedings, it will often prove difficult (in some cases impossible) to investigate, prepare or present the claim properly, though the reason for this may be masked over time.

Following the initial consultation (which should provide ample basis for a draft witness statement), a case plan or agreed list of actions should be settled upon. In the early stages of a claim, the nature and extent of the client's injury permitting, a collaborative approach to obtaining evidence is often highly productive. The essential evidential considerations are as follows:

- documentary evidence;
- photographic and video evidence;
- witness evidence;
- medical evidence;
- expert evidence.

12.3 DOCUMENTARY EVIDENCE

The client will often be able to provide assistance with regard to the employment history and as to any losses suffered but, for the most part, the task of obtaining documentary evidence will rest with the legal adviser by way of disclosure.

12.3.1 Third Party Disclosure

The proper approach to documentary requests is set out at para 2.12 of the Pre-Action Protocol:

> 'The aim of the early disclosure of documents by the defendant is not to encourage "fishing expeditions" by the claimant, but to promote an early exchange of relevant information to help in clarifying or resolving issues in

> dispute. The claimant's solicitor can assist by identifying in the letter of claim or in a subsequent letter the particular categories of documents which they consider are relevant.'

The employer's obligation to provide documentary evidence is broadly framed with reference to the 'standard disclosure' obligations arising under CPR 31.6. The Pre-Action Protocol makes specific provision regarding documents at paras 3.10–3.12 as follows:

- If the defendant denies liability, he should enclose with the letter of reply documents in his possession that are material to the issues between the parties, and that would be likely to be ordered to be disclosed by the court, either on an application for pre-action disclosure, or on disclosure during proceedings.

- Attached to Annex B of the Pre-Action Protocol are specimen, but non-exhaustive, lists of documents likely to be material in different types of claim. Where the claimant's investigation of the case is well advanced, the letter of claim could indicate which classes of documents are considered relevant for early disclosure. Alternatively these could be identified at a later stage.

- Where the defendant admits primary liability, but alleges contributory negligence by the claimant, the defendant should give reasons supporting those allegations and disclose those documents from Annex B which are relevant to the issues in dispute. The claimant should respond to the allegations of contributory negligence before proceedings are issued.

- No charge will be made for providing copy documents under the protocol.

Annex B to the Pre-Action Protocol contains a generic list of documents likely to be applicable in all 'workplace claims':

- accident book entry;
- first aid report;
- surgery record;
- foreman/supervisor accident report;
- safety representative's accident report;
- RIDDOR report to HSE;
- other communications between defendant and HSE;

- minutes of health and safety committee meeting(s) where accident or matter considered;
- report to DSS;
- documents listed above relative to any previous accident or matter identified by the claimant and relied upon as proof of negligence;
- earnings information where defendant is employer;
- documents produced to comply with requirements of the Management of Health and Safety at Work Regulations 1999.[3]

Within the following section of Annex B, specimen lists are provided where specific pieces of health and safety legislation applies. Advisers should note that in some instances the legislative provisions may have been superseded and, in any event, that the lists are 'non-exhaustive'.

For example, additional documents may be relevant where the Provision and Use of Work Equipment Regulations 1998[4] apply and an appropriately tailored request may properly include documents that an employer might have been expected to create in compliance with the Work at Height Regulations 2005,[5] which are not presently listed.

In each case, the disclosure request should be very carefully considered, so as to encompass all relevant documents and to exclude those that are not applicable in the circumstances. Blanket requests attract criticism and must be avoided at all costs. The request should instead be seen as an opportunity to demonstrate careful preparation and a clear grasp of the relevant issues.

In cases where medical causation is likely to be an issue (in manual handling and other straining injury cases, for example) or where long-term absence from work may be a feature, all personal information retained by the relevant employer should also be requested (with provision of an appropriate mandate) at an early stage, eg personnel and occupational health records.

12.3.2 Other Sources

In the context of accidents at work, the amount of useful documentary evidence readily available to advisers using the internet and other sources is astonishing.

---

3 SI 1999/3242.
4 SI 1998/2306.
5 SI 2005/735.

Where a relevant approved code of practice ('ACOP') has been issued by the HSC, it has been authoritatively stated that the courts should consider such guidance when construing the meaning of any legislative provision.[6]

In a wider sense, advisors should always consider whether a relevant ACOP or other official guidance document assists in determining the adequacy of any risk assessment (whether in a general or specific context) or issues such as 'suitability' and 'training' relating to equipment supplied pursuant to PUWER[7] or the PPE Regulations,[8] for example.

The HSE also publish a wide range of industry and task specific guidance documents, many of which can be downloaded without cost.[9] It may also be worth researching whether any applicable union[10] (with an employee focus) or trade association[11] (with an employer focus) has published documentary guidance concerning issues of health and safety in a particular context.

When considering claims relating to equipment (unsuitability, use or a simple failure to provide the tools necessary for the task in hand), documentation from manufacturers and suppliers can assist greatly, where requested. For example, something as simple as a picture of a cheap, collapsible sack trolley may well destroy an employer's defence of 'reasonable practicability' in a manual handling claim. Equally, brochures or other product information may assist greatly in a PPE claim, as evidence in rebuttal as to 'suitability'.

Reports to confirm scientific or technical information are also readily available over the internet. For example, the Met Office can provide historical weather data, where the conditions in which an accident occurred are relevant,[12] while the Royal Observatory (Edinburgh) can confirm sunrise and sunset times on a particular day, in a particular location.[13]

---

6 See dicta of Smith LJ in *Ellis v Bristol City Council* [2007] EWCA Civ 685,CA, at [32]–[33]

7 Provision and Use of Work Equipment Regulations 1998, SI 1998/2306.

8 Personal Protective Equipment at Work Regulations 1992, SI 1992/2966.

9 Visit www.hse.gov.uk/pubns/index.htm.

10 The TUC maintain a list of employee groups (www.tuc.org.uk/tuc/unions_main.cfm). It is often worth searching individual websites for documents, case studies and awareness campaigns.

11 The TAF maintain a list of trade groups: (www.taforum.org/searchgroup.pl?sector=1;n=500). It is often possible to order or download industry specific guidance on health and safety.

12 Visit www.metoffice.gov.uk/legal.

13 Visit www.roe.ac.uk/info/srss.html.

12.4 PHOTOGRAPHIC AND VIDEO EVIDENCE

It is trite to say that 'a picture tells a thousand words', but the cliché is never more aptly used than in the context of an accident at work claim. Though the problem is not an uncommon one,[14] it is unacceptable for claims to fail simply because the trial judge is unable to reach the necessary findings of fact as to the mechanism or other relevant circumstances of an accident.

In many cases, a few photographs of the relevant location and any relevant equipment will suffice. In any case where the employee's position, actions or interrelationship with equipment (or the cycle or limits of operation of the equipment itself) are relevant, video evidence should be obtained.

Photographic or video evidence must be sufficiently clear and provide a view of the surrounding area (to place the equipment or position of the injured party in context). Injury permitting, the employee should attend any site inspection and demonstrate his or her position and actions.

12.5 WITNESS EVIDENCE

The drafting of witness statements in accident at work claims is unfortunately one of the most neglected areas. The importance of good witness statements cannot be understated.

In addition to the statement of case, the principal witness statements are the main (often the only other) documents considered by a trial judge prior to the commencement of a final hearing. If the witness statements do not effectively present the claim or fail to address key aspects of the necessary evidence, it is difficult (often impossible) to salvage matters at the eleventh hour.

Preferably draft statements should be reviewed in conference with intended trial counsel prior to commencement of proceedings, where disclosure has taken place under the Pre-Action Protocol, and in any event prior to formal exchange of lay witness evidence. The courts typically expect and encourage such input in fast track and multi track proceedings alike.

12.5.1 Formalities

While honoured more in the breach than in the observance, para 17 of CPR 32 PD provides a number of specific requirements regarding the appropriate heading of witness statements for reference purposes.

---

[14] See, for example, the recent case of *Verlander v Devon Waste Management* [2007] EWCA Civ 835, CA.

Paragraph 18 of CPR 32 PD provides more purposeful advice regarding content. The witness statement must, if practicable, be in the intended witness's 'own words' rather than the lawyers. The statement should be expressed in the first person and must state the following information:

- the full name of the witness;
- his place of residence or, if he is making the statement in his professional, business or other occupational capacity, the address at which he works, the position he holds and the name of his firm or employer;
- his occupation, or if he has none, his description; and
- the fact that he is a party to the proceedings or is the employee of such a party if that is the case.

Paragraph 18 also makes specific provision for a witness to identify the basis or source of the statement's content and for the exhibiting and identification of any other evidence relied upon by the deponent.

While para 19.1 of CPR 32 PD makes generic provision regarding the format of all witness statements, the guidance offered in para 19.2 is a helpful reminder in the specific context of accident at work claims:

> 'It is usually convenient for a witness statement to follow the chronological sequence of the events or matters dealt with, each paragraph of a witness statement should as far as possible be confined to a distinct portion of the subject.'

It can often be a difficult and time-consuming exercise for a trial judge to fully assimilate all of the factual information relevant for the purposes of making a fair determination. That task is made more complicated (and less attractive) if the witness statements do not follow a logical structure and incorporate large rather than 'bite-sized' pieces of a narrative.

All witness statements must, of course, be accompanied by a signed 'statement of truth' by the deponent, pursuant to CPR 22.1.

12.5.2 Setting the Scene

Witness statements must start with a clear and comprehensive introduction. If the injured party's employment history is relevant, details should be provided at the beginning. Particulars of any relevant qualification or training should be indicated and explained, with copies of any certificates exhibited.

The injured party should also explain his capacity, job description and working practices with an appropriate level of detail. The working environment and any relevant features or work equipment should be described, similarly with all relevant photographs exhibited.

It is often helpful to introduce any relevant people at an early stage (whether of not the court is to receive their evidence directly). These may, for example, include co-employees, a site supervisor or manager, a first aider or union representative.

12.5.3 The Accident Circumstances

With the scene appropriately set, it will hopefully be possible to briefly state the mechanism or precise circumstances of the accident. Avoid overly complicated descriptions and, above all, any element of conjecture. If there are gaps in the employee's recollection or understanding as to precisely what happened, an appropriate concession is preferable to any attempt, however innocent, to fill evidential gaps.

If there had been anything different about the manner in which the injured party had been working at the time of the accident, when compared with his or her usual role or practices, precise details and an explanation regarding the reasons or circumstances for such a change should be provided.

This part of the witness evidence must, where possible, clearly indicate how the relevant injury occurred and, particularly in the context of back and other straining injuries, provide a precise description regarding the onset of discomfort.

12.5.4 Consequential Matters

Whilst avoiding consideration of quantum here, the statement must provide details of the accident's physical effects (was the injured party able to continue or resume work?), any immediate medical assistance received (by a first aider, ambulance, walk-in-centre or hospital etc) and if, when and how the accident was reported to the employer.

The injured party (and any co-employee or union/safety representative who provides witness evidence) must supply details regarding any post-accident consultation, training, instruction or information. If changes to the working environment, equipment or procedures are introduced (whether or not this is attributed to the accident) this must also be fully particularised.

Whilst avoiding adverse comment or detailed allegations of breach of duty, it is appropriate for the injured party (and any other relevant witness) to identify key documents and suggest how the accident might have been avoided.

12.6 MEDICAL EVIDENCE

For the reasons set out more fully at start of this chapter, careful selection and instruction of medical experts is vital. Medical causation is becoming an increasingly contentious issue, particularly in manual handling cases and/or where the injured person has a relevant medical history.

Paragraphs 2.14 and 2.15 of the Pre-Action Protocol provide some general guidance:

> 'The protocol encourages joint selection of, and access to, experts. The report produced is not a joint report for the purposes of CPR Part 35. Most frequently this will apply to the medical expert, but on occasions also to liability experts, eg engineers. The protocol promotes the practice of the claimant obtaining a medical report, disclosing it to the defendant who then asks questions and/or agrees it and does not obtain his own report. The Protocol provides for nomination of the expert by the claimant in personal injury claims because of the early stage of the proceedings and the particular nature of such claims. If proceedings have to be issued, a medical report must be attached to these proceedings. However, if necessary after proceedings have commenced and with the permission of the court, the parties may obtain further expert reports. It would be for the court to decide whether the costs of more than one expert's report should be recoverable.
>
> Some solicitors choose to obtain medical reports through medical agencies, rather than directly from a specific doctor or hospital. The defendant's prior consent to the action should be sought and, if the defendant so requests, the agency should be asked to provide in advance the names of the doctor(s) whom they are considering instructing.'

While the use of medical agencies arguably has its place in some areas of personal injury practice, it is preferable for medical records to be obtained directly by the legal adviser in all accident at work claims and for the medical expert to be selected and instructed on a case-by-case basis.

Medical records should ideally be requested as soon as possible, assuming a positive risk assessment has been made during the initial consultation. Delays in processing requests are an increasing problem and can significantly delay the progress of a claim.

While it is obviously not appropriate to 'coach' clients prior to a medico-legal examination, it is sensible and proper for advisers to review medical records, confirm any relevant details (accident and initial treatment dates etc), take detailed instructions on the treatment chronology and ask about any potentially relevant entries within the past medical history.

Advisers must remember that medico-legal examinations are typically a highly unusual, often daunting experience. Inconsistencies or omissions on the part of the client during the consultation, however innocent, often influence the expert's opinion and will certainly be seized upon the third party's representatives. In most cases, such problems are avoidable if the process and implications of the examination are fully explained.

In terms of expert selection, a wealth of assistance is readily available. APIL maintains a comprehensive database of experts across a broad range of specialist medical and non-medical areas (reassuringly, the database is drawn from member recommendations, unless otherwise indicated). The experiences of others within the practice and of local, specialist counsel should also be sought. In cases where the instruction is to be made on a unilateral basis, it may be desirable (particularly in respect of less commonly arising injuries) to make direct enquiries of a proposed expert to accurately determine his or her experience.

12.7 EXPERT EVIDENCE

The issue of whether (non-medical) expert evidence is required in an accident at work must be approached cautiously. The CJC has produced a protocol concerning the instruction of experts in all civil proceedings.[15] Paragraph 6 of the protocol 'The Need for Experts' provides as follows:

- Those intending to instruct experts to give or prepare evidence for the purpose of civil proceedings should consider whether expert evidence is appropriate, taking account of the principles set out in CPR Parts 1 and 35, and in particular whether:
  (a) it is relevant to a matter that is in dispute between the parties;
  (b) it is reasonably required to resolve the proceedings (CPR 35.1);
  (c) the expert has expertise relevant to the issue on which an opinion is sought;
  (d) the expert has the experience, expertise and training appropriate to the value, complexity and importance of the case; and
  (e) these objects can be achieved by the appointment of a single joint expert.

- Although the court's permission is not generally required to instruct an expert, the court's permission is required before experts can be called to give evidence or their evidence can be put in (CPR 35.4).

---

15 *Protocol for the Instruction of Experts to give Evidence in Civil Claims* (June 2005) as annexed to CPR Part 35.

In the overwhelming majority of accident at work claims, the court will be reluctant to accept that expert evidence is reasonably required to resolve the proceedings. Advisers should ask themselves whether the trial judge will be in a position to test the validity of parties' respective arguments on an issue of complexity without expert assistance. Leave applications should be presented on the same basis.

'Equality of arms' is often a relevant issue. If the particular employer has senior, professionally qualified operatives or other relevant expertise within its undertaking, the court may be more inclined to accede to an application for expert evidence as to liability issues.

Examples of cases in which it is more common for the court to grant leave to call a liability expert include the following:

- where the employer (in discharge of its statutory responsibilities) has sought professional assistance;
- where ergonomics or other specialist scientific knowledge is a feature;
- where engineering input is required (e g for inspection of machinery or to determine whether engineering solutions could and should have employed to reduce a relevant risk);
- in complex lifting and handling cases, for example in the context of nursing care and restraint of violent individuals.

The guidance provided above, urging care regarding the selection and instruction of medical experts, is equally applicable in this context. The right expert, properly appointed, can make or break a claim.

CHAPTER 13

PRACTICE AND PROCEDURE

13.1 INTRODUCTION

If properly prepared and presented, most accident at work claims should be capable of resolution by agreement. To reach this point, however, it will often be necessary to take a number of pre-issue steps to investigate and preserve the claimant's position.

If a fair settlement cannot be reached, it is imperative that advisers make full, effective use of the CPR to maximise the prospects of a successful outcome for the injured person.

Whilst there are a number of comprehensive, helpful guides to civil procedure in general and personal injury litigation in particular, there are a number of discrete, practical issues that may arise in the context of an accident at work claim. The following guidance is not intended to be exhaustive, but rather to promote tactical awareness and to provide a useful starting point for advisers.

13.2 PRE-ISSUE APPLICATIONS AND URGENT ACTION

13.2.1 Disclosure by Intended Parties

Obtaining early, full disclosure is central to the effective assessment and prosecution of all accident at work claims. In almost any employment context numerous health and safety duties will apply and the courts will expect to see a clear (in some cases, voluminous) paper trail before accepting proper compliance on the part of the employer.

Where a full or partial denial of liability is made, the proposed defendant is expected to make 'standard disclosure' under the pre-action protocol, which includes the following material:[1]

- any document upon which the party intends to rely;

---

[1] As defined by CPR 31.6.

- any document that adversely affect the party's own case;
- any document that adversely affect another party's case;
- any document that supports another party's case; and
- any document which the party is required to disclose pursuant to a relevant practice direction.

The proposed defendant must provide standard disclosure immediately upon making a denial of liability and/or upon raising any allegation of contributory negligence.

If disclosure is not forthcoming or the material provided is incomplete, the next step for an adviser is to identify the alleged failure(s) and to put the proposed defendant on notice that a court application will be made in the event of continued default. Naturally, if a reasoned explanation for delay is offered by the defendant or its insurer, a sensible approach will need to be adopted.

In the absence of an acceptable response, a short period of further grace should be indicated in correspondence (it need not be longer than 14 days from anticipated receipt of the letter[2]), with notice that the costs of any application will be sought in default. Protracted exchanges of correspondence must be avoided. These lead to unnecessary, potentially prejudicial delay and there is a growing tendency for judges to limit costs recovery in this context.

The procedure for making an application for disclosure prior to the start of proceedings is governed by CPR 31.16.[3] Applications must be supported by evidence, outlining the following matters:

- a brief description of the dispute (in most cases, no more than the information provided within the letter of claim);
- a short chronology, cross-referenced to the anticipated timetable under the pre-action protocol, with all relevant correspondence exhibited; and
- a list of the documents required (it will suffice to refer to the list provided within the draft order accompanying the application) and a

---

[2] The minimum period prescribed for notice of a summary judgment hearing (and thus ultimate disposal of proceedings) is 14 days by comparison. The proposed defendant in this context will already have had three months within which to respond. Fourteen days is certainly sufficient time within which to make some form of substantive response, if only an explanation for the delay.

[3] The Court's jurisdiction, founded under s 33 of the Supreme Court Act 1981 and s 52 of the County Courts Act 1984, originally arose in the context of personal injury claims, but now extends to all civil proceedings.

brief statement (or explanation if the documentation is not included within the annex B to the protocol) as to their relevance.

The powers of the court under CPR 31.16 are circumscribed, albeit the conditions are commonly met in most accident at work claims. An order can only be made where[4]:

- the respondent is likely to be a party to subsequent proceedings;
- the applicant is also likely to be a party to those proceedings;
- if proceedings had started, the respondent's duty by way of standard disclosure would extend to the documents or classes of documents of which the applicant seeks disclosure; and
- disclosure before proceedings have started is desirable in order to:
  (i) dispose fairly of the anticipated proceedings;
  (ii) assist the dispute to be resolved without proceedings; or
  (iii) save costs.

An application under CPR 31.16, as with all others, must be accompanied by a draft order, which should incorporate the following[5]:

- a list specifying the documents or the classes of documents that the respondent must disclose;
- a requirement, when making disclosure, for the respondent to specify any of those documents
  (i) that are no longer in its control; or
  (ii) in respect of which a right or duty to withhold inspection is claimed;
- a requirement that the respondent indicate what has happened to any documents that are no longer in its control; and
- a time limit for disclosure (and inspection if appropriate).

The general principles applicable to CPR.36.16 were considered in *Black v Sumitomo Corpn.*[6] 'Likely to be a party' means 'might well' be so if proceedings are issued.

The higher courts consistently encourage a pragmatic approach.[7] Judges should consider whether disclosure will help the parties to focus more

---

4 CPR 31.16(3).

5 The first two limbs are mandatory under CPR 31.16(4), whilst the second two limbs are discretionary under CPR 31.16(5), but no less important in this context.

6 [2001] EWCA Civ 1819.

7 See *Marshall & Marshall v Allotts* [2004] EWHC 1964, QB.

clearly on the relevant issues and in their eventual statements of case, should an order be made. Failure to make a proper response to the pre-action protocol is always a prominent consideration.

The courts will typically make provision for costs recovery on a standard basis against the respondent, because of default in compliance with the pre-action protocol.

Full or partial exceptions may be made where, for example, the claimant's representative has prevaricated or the extent of the disclosure request was excessive. Further to the advice offered in Chapter 12 above, this exercise should be seen as an opportunity to demonstrate careful preparation and a clear grasp of the relevant issues, and thus to prompt early admissions.

13.2.2 Disclosure by Non-Parties

The court may order a person who is not an intended party to provide disclosure within existing proceedings, by virtue of CPR 31.17, and exceptionally prior to commencement of a claim, where this is necessary in all the circumstances. Prior to commencement of proceedings, the jurisdiction of the court is closely curtailed. An order can only be made, assuming it is otherwise just to do so, where each of the following conditions is met[8]:

- a wrong must have been carried out, or arguably carried out, by an ultimate wrongdoer;
- there must be the need for an order to enable action to be brought against the ultimate wrongdoer (ie the required information cannot be sought elsewhere or on an alternate basis, eg pursuant to CPR 31.16); and
- the person against whom the order is sought must:
  - (a) be mixed up in so as to have facilitated the wrongdoing; and
  - (b) be able or likely to be able to provide the information necessary to enable the ultimate wrongdoer to be sued.

While typically the preserve of complex commercial or defamation claims, this type of order may exceptionally be appropriate in the context of an accident at work claim. In any such case, the specialist advice of an experienced litigator should be sought.

---

8 *Mitsui & Co Ltd v Nexen Petroleum UK Ltd* [2005] EWHC 625 (Ch), [2005] 3 All ER 511.

13.2.3 Preservation of Evidence and Property

Workplaces and organisations undergo constant change, often to the ultimate benefit of employees in terms of improvements to health and safety. This will, however, prove 'cold comfort' to an injured person whose claim is made more difficult to present, for example owing to changes to plant and equipment or the loss of documentation following insolvency or restructuring.

Among the court's powers to grant interim remedies, pursuant to CPR Part 25, specific provision is made for making orders in the following terms:

- for the detention, custody or preservation of relevant property;
- for the inspection of relevant property;
- for the taking of a sample of relevant property;
- for the carrying out of an experiment on or with relevant property; and
- to authorise a person to enter any land or building in the possession of a party to the proceedings (or likely party to anticipated proceedings) for the above purposes.

The term 'relevant property' means 'property (including land) which is the subject of a claim or as to which any question may arise on a claim' and thus includes a workplace or equipment about which a relevant allegation is made or which provides context to the happening of an accident.

13.2.4 Limitation

Advisors must always be alert to the issue of limitation upon receipt of initial instructions. Prospective clients who wait a couple of years or more before seeking legal assistance may well have become confused about dates with the passage of time.

In the first instance, a point of reference should be established (family birthday, event or public holiday etc). Prompt consideration should be given to issuing a claim form, while further investigations are conducted, in any case where there is concern that the end of the primary limitation period may be approaching (or have already passed).

Whilst less commonly an issue in the context of accidents at work (when compared, for example, with industrial disease claims) advisers should note the four key requirements necessary to establish a person's date of knowledge (the point at which time begins to run) for the purposes of s 11 and 14 of the Limitation Act 1980, namely:

- that the injury complained of was significant;
- that the injury was attributable in whole or in part to the act or omission that is alleged to constitute negligence or breach of duty;
- the identity of the defendant; and
- if it is alleged that the act or omission was that of a person other than the defendant, the identity of that person and the additional facts supporting the bringing of an action against the defendant

Most employers' liability claims arise in the context of 'discrete events' where both the significance of the injury and the acts or omissions that gave rise to it can be readily appreciated by the injured person. The first two limbs of the above test must, however, be carefully considered in the context of injury arising out of a process (not necessarily an industrial disease but, for example, a lumbar or cervical spine complaint associated with repetitive manual handling activities[9]) or where symptoms have developed over time.

The second two limbs of the test may well be relevant where there is genuine ignorance or confusion as to the precise structure of an employer organisation or its relationship with third parties.

In *Cressey v E Timm & Son Ltd*,[10] the employer unsuccessfully appealed (twice) against the finding that a claim form had been issued within three years of the employee's 'date of knowledge' for the purposes of the Limitation Act 1980. Mr Cressey had been misinformed as at the date of his accident, without any fault on his part, about the identity of his employer and had only discovered it later (less than three years prior to commencement of proceedings). Providing the judgment of the Court, Rix LJ observed as follows:[11]

> '...It is likely that in most cases of an accident at work, the employee will there and then have knowledge of the identity of his employer, and therefore of the defendant. However, in a minority of cases, where the identity of the employer is uncertain, as in Simpson, or even wrongly stated to the employee, as here, the date of knowledge may well be postponed. How long it will be postponed by will depend on the facts of such cases. In general I do not believe that it can be postponed for long: only as long as it reasonably takes to make and complete the appropriate enquiries. But if such enquiries are met by misinformation, or a dilatory response, again as in Simpson, then it is not possible to be dogmatic about the right conclusion ....'

[9] See, for example, *Knott v Newham Healthcare NHS Trust* [2002] EWHC 2091, QB.
[10] [2005] EWCA Civ 763, (2006) ICR 282, (2006) PIQR P9.
[11] [2005] EWCA Civ 763, (2006) ICR 282, (2006) PIQR P9 at [28].

13.3 PRACTICALITIES UPON ISSUE AND CHOICE OF DEFENDANT

13.3.1 Identity of Employer and/or Insurer

The case of *Cressey* above aptly illustrates the difficulties that identity can present in an otherwise straightforward case. Identifying the correct name of an employer is normally a simple enquiry to conduct. Prior to issue and service of proceedings, in the absence of a nominated solicitor, it is imperative that advisers confirm such details (through the insurer, directly with the undertaking and/or using a company search or agent as appropriate).

In other accident at work claims, particularly those involving agency workers, the self-employed and mobile workers, the exercise of identifying the correct defendant and/or its liability insurer can be far more complicated (in some cases impossible).

Budding detective work is often the order of the day. This may extend to enquiries of co-employees, representatives, managers or directors, local occupiers or businesses. APIL also hold a database of insurers and past information requests by members. In more substantial cases, where the necessary resources can be justified, a detailed insurance archaeology may be required.[12]

In exceptional cases, as suggested above, the use of non-party disclosure applications may be appropriate in this context.

13.3.2 Multiple Defendants

Cases involving two or more potential defendants require particular care. The main issue here is one of costs – the twofold risk that in the event of pursuing two or more defendants, but only succeeding against one, recovery of costs may be limited (i.e. restricted only to those costs incurred in pursuit of the unsuccessful defendant) and that the injured party may be ordered to pay the costs of the successful defendant.

Some helpful guidance is provided in *Moon v Garrett*.[13] Mr Moon sustained injury when, in the process of delivering concrete blocks to Mr Garrett's premises, he fell and rolled into a pit. Mr Moon brought his claim against Mr Garrett, as occupier, and also against his own employer. Only the first claim succeeded, but the trial judge directed that Mr Garrett

---

[12] For simple yet detailed guidance, see Nick David 'How to conduct an insurance archaeology: a layman's guide' (2004) 14(1) PI Focus.

[13] [2006] EWCA Civ 1121, (2007) ICR 95, (2007) PIQR P3.

should pay the costs of both Mr Moon and his employer.[14] Dismissing Mr Garrett's appeal on this issue, Waller LJ observed as follows:

> '... It seems to me ... that there are no hard and fast rules as to when it is appropriate to make a Bullock or Sanderson order. The court takes into account the fact that, if a claimant has behaved reasonably in suing two defendants, it will be harsh if he ends up paying the costs of the defendant against whom he has not succeeded. Equally, if it was not reasonable to join one defendant because the cause of action was practically unsustainable, it would be unjust to make a co-defendant pay those defendant's costs. Those costs should be paid by a claimant. It will always be a factor whether one defendant has sought to blame another.
>
> The fact that cases are in the alternative so far as they are made against two defendants will be material, but the fact that claims were not truly alternative does not mean that the court does not have the power to order one defendant to pay the costs of another. The question of who should pay whose costs is peculiarly one for the discretion of the trial judge ....'

Advisors should be careful to only maintain claims against multiple defendants where it realistic to do so – i e where relevant health and safety duties operate co-extensively (or at least where this is arguably so) and/or where separate duties were owed to the injured person.

Prior to issue and service of such proceedings, it is prudent to write to each proposed defendant in like terms, briefly outlining all relevant allegations and inviting an open admission of liability or (at least) agreement as to the issue apportionment. Such correspondence might then be relied upon should the issue of costs ultimately fall to be determined by the trial judge.

13.3.3 Addition and Substitution

If the wrong party is named or another potential defendant is identified within the proceedings, prompt consideration must be given to the making of an application under CPR Part 19. Prior to the expiry of limitation, the court may order as follows:[15]

- a person to be added as a new party if –
  (a) it is desirable to add the new party so that the court can resolve all the matters in dispute in the proceedings; or

---

14 A *Sanderson* order – *Sanderson v Blyth Theatre Co* [1903] 2 KB 533, CA. Note also the availability of a *Bullock* order in this context, requiring the claimant to pay the successful defendant's costs, but allowing these to be recovered as part of the claimant's costs of the action: *Bullock v London General Omnibus Co* [1907] 1 KB 264, CA.

15 CPR 19.2.

(b) there is an issue involving the new party and an existing party that is connected to the matters in dispute in the proceedings, and it is desirable to add the new party so that the court can resolve that issue.

- any person to cease to be a party if it is not desirable for that person to be a party to the proceedings;

- a new party to be substituted for an existing one if –
  (a) the existing party's interest or liability has passed to the new party; and
  (b) it is desirable to substitute the new party so that the court can resolve the matters in dispute in the proceedings.

Where the primary limitation period has expired, the powers of the court are more closely circumscribed.[16] An order for addition or substitution can only be made if:

- the relevant limitation period was current when the proceedings were started; and

- the addition or substitution is necessary.

The addition or substitution of a party will only be considered 'necessary' in this context if the court is satisfied that:

- the new party is to be substituted for a party who was named in the claim form in mistake for the new party;

- the claim cannot properly be carried on by or against the original party unless the new party is added or substituted as claimant or defendant; or

- the original party has died or had a bankruptcy order made against him and his interest or liability has passed to the new party.

13.3.4 Insolvency

Company employers that fail to implement proper health and safety in the workplace are often those which also end up in financial difficulties.

In the event of impending insolvency, a number of practical steps may need to be taken, depending upon the company's situation.[17] In each case, it is prudent to seek specialist advice from a licensed insolvency practitioner, other qualified adviser or counsel (ordinarily recoverable as a proper disbursement within the proceedings).

[16] CPR 19.5.

[17] See Ian Pears 'The Insolvent Defendant Company' (2003) 13(2) PI Focus.

In the event of dissolution (the company ceasing to exist), an application will need to be made to restore the company to the register.[18] Similarly, it is prudent to seek specialist advice in this context.

Once judgment has been obtained, the injured party steps into the shoes of the insolvent company and can claim against any relevant insurance policy pursuant to the Third Party (Rights Against Insurers) Act 1930.

[18] See the Treasury Solicitor's 'Guide to Company Restoration and Dissolution Void Applications' (www.tsol.gov.uk/Publications/companyrest.pdf).

INDEX

References are to paragraph numbers.